Caring for Llamas & Alpacas:

A Health and Management Guide

by

Clare Hoffman, DVM and Ingrid Asmus

Illustrated by Ingrid Asmus

Rocky Mountain Llama and Alpaca Association, Inc.

Library of Congress Number: 89-61100

ISBN 0-9622768-0-4

First Printing: May, 1989
Second Printing: April, 1990
Third Printing: November, 1991
Fourth Printing: April, 1993
Fifth Printing: October, 1994
Second Edition, First Printing: March, 1996
Second Edition, Second Printing: March, 1998
Second Edition, Third Printing: November, 2000
Second Edition, Fourth Printing: March, 2003
Second Edition, Fifth Printing: October, 2005

Printed in the United States of America by: Pioneer Printing, Cheyenne, WY

Cover Photograph: Llamas and Alpacas provided by Bobra Goldsmith
                     Rocky Mountain Llamas - Longmont, Colorado

Published by: Rocky Mountain Llama and Alpaca Association, Inc.

Distributed by: RMLA Bookstore   Phone: 303-621-2960
                     Attn: Janice Adamcyk
                     39420 Olson Court
                     Kiowa, CO 80117-9604
                     Email: adamcyk@earthlink.net

For RMLA Membership Information contact:
                     Barbara & Bob Hance, Membership Committee
                     11818 West 52nd Avenue
                     Wheat Ridge, CO 80033
                     303-422-4681
                     Email: hancelama@att.net

Visit our web site: www.RMLA.com

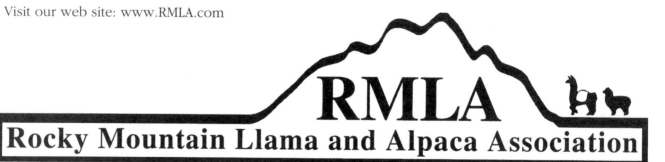

©2005 RMLA

CONTENTS

ILLUSTRATIONS

ACKNOWLEDGMENTS

Our many thanks go out to Dr. LaRue Johnson and Bobra Goldsmith for the multitude of information gleaned. A special thanks to Dr. LaRue Johnson, Howard Kerstetter, Bobra Goldsmith and Brad Sprouse for their help in editing this book and to Judy Brock for putting it all on disc.

In addition, our thanks go to all of the llama owners who helped and supported us as we learned about llamas, and to all of the llama and alpaca owners that may find this book a helpful addition to their llama library.

Clare Hoffman, D.V.M.
Ingrid Asmus

PREFACE

The authors and RMLA have attempted to provide you with a book which will assist you in the care of your llamas or alpacas. The proper application of any medical procedure must necessarily be prefaced with an informed diagnosis. In each case, symptoms may vary significantly, and reasonable minds may thus differ as to both the diagnosis and the treatment.

Thus, our reasonable readers will understand why the authors, RMLA and the sponsors of *Caring for Llamas and Alpacas: A Health and Management Guide* are expressly NOT WARRANTING OR RECOMMENDING ANY PROCEDURES OR PRODUCTS described in this text. Occasionally, brand names are used as examples of products, but this does not mean that similar products made by other companies should not be used. This book is not intended to be a substitute for a llama and alpaca-trained doctor of veterinary medicine and will certainly not turn the reader into one. Use the procedures and products at your own risk, as the Rocky Mountain Llama and Alpaca Association, Inc. cannot be responsible for the results that may be obtained. Any and all damages are limited to a refund of the purchase price of the text.

PUBLISHER'S ACKNOWLEDGEMENTS

The Rocky Mountain Llama and Alpaca Association, Inc. was founded in 1982 to provide education and information about llamas and share the enjoyment of these special animals. The publication of this book is in keeping with the Association's objectives, and we hope it will be a useful contribution to the entire llama and alpaca community.

RMLA was the first llama organization to actively support llama research on a continuing basis, and we are pleased that RMLA will continue this support by dedicating a portion of the proceeds from *Caring for Llamas and Alpacas* to research.

We wish to thank Clare Hoffman and Ingrid Asmus for recognizing the need for a basic reference on llama care and doing something about it. We deeply appreciate their many hours of writing and drawing.

Lastly, publication would not have been possible without financial backing. RMLA gratefully acknowledges the following RMLA members for their special support:

LLAMAS OF BRIDGER LANE
Betty and Bob Barkdoll
Douglas, WY

SUKI LLAMAS
Suki Dewey
Boulder, CO

BLUE SKYE LLAMAS
Michele and Bruce Fickel
Berthoud, CO

HAST LLAMAS
Erma and Bernie Hast
Montrose, CO

STORM KING LLAMAS
Mark Lee and Gregg Velasquez
Silt, CO

McROBERTS LLAMAS
Barbara and Jerry McRoberts
Kathy and Wayne McRoberts
Gurley, NE

OVERLAND TRAIL LLAMAS
Dale and Mike Pettigrew
Livermore, CO

WESTERN SLOPE LLAMAS
Bob and Nancy Russell
Salmon, ID

ROCKY MOUNTAIN LLAMAS
Bobra and Ulo Goldsmith
Longmont, CO

TIMBERLINE LLAMAS
Wes and Mary Mauz
Golden, CO

SWITZER-LAND ALPACAS AND LLAMAS
Chris and Phil Switzer
Estes Park, CO

PFOUR PFRANGLE LLAMA PFARM
Kirby and Louie Pfrangle
Worland, WY

WILDIERNESS WAY LLAMAS
Jodi and Don Sleeper
Ocate, NM

HIGH LLAMA
Nancy and Cutler Umbach
McCall, ID

TEJAS LLAMAS
Andrea and David Wieboldt
Coupland, TX

INDIAN PEAKS LLAMAS
Alaine Byers
Ward, CO

GREAT DIVIDE LLAMAS
Stan and Diane Ebel
Loveland, CO

SILVER SAGE LLAMAS
Dan and Ellen Schreiner
Casper, WY

COLUMBINE LLAMAS
Wally White
Durango, CO

WIND RIVER LLAMAS
Anne Johnson and
Darlene Vaughan
Lander, WY

PARTS OF A LLAMA

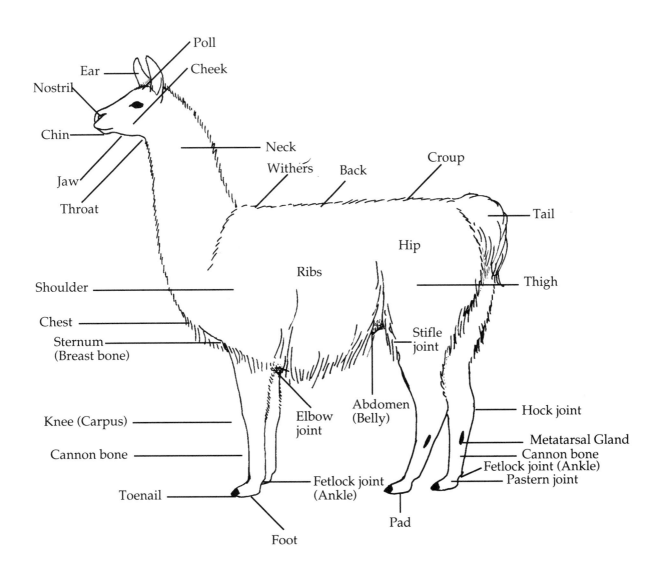

Poll

Ear

Cheek

Nostril

Chin

Jaw

Throat

Neck

Withers

Back

Croup

Tail

Hip

Ribs

Thigh

Shoulder

Chest

Sternum
(Breast bone)

Stifle
joint

Hock joint

Metatarsal Gland

Knee (Carpus)

Cannon bone

Elbow
joint

Abdomen
(Belly)

Cannon bone

Fetlock joint (Ankle)

Pastern joint

Toenail

Fetlock joint
(Ankle)

Pad

Foot

INTRODUCTION

Llamas are becoming increasingly popular in the United States. Because they are somewhat "exotic" animals, it has been difficult to find basic information about them. This gap is being filled gradually by a number of publications. However, the available information on health and husbandry of llamas is sparse. In addition, many first-time llama owners have little or no experience in raising large animals. This can result in difficulty noticing problems or indications of the llama's health status. Even experienced livestock raisers may find themselves at a loss in understanding or caring for llamas, which are different from other domestic animals.

This book is designed to serve as a guide to health care for the LLAMA OWNER. It is NOT meant to replace the need for a veterinarian, but rather to alert you as to WHEN veterinary assistance is needed. This book will help you distinguish between cases that can just be watched for a while, versus cases that require immediate veterinary attention. It also contains first aid tips to care for an injured or ill llama when out on the trail or until a veterinarian can be reached.

The chapters dealing with emergency health situations contain a troubleshooting chart. This enables the reader to look up a particular problem quickly to see whether the situation is indeed an emergency and, if it is, what immediate steps should be taken.

Then, after the crisis subsides, you may leisurely read through the explanation of the problem in the text. A thorough index is included for this very reason. We recommend that you become thoroughly familiar with the format of the book.

The information we chose to cover in this book is based on questions commonly asked by llama owners and on the most common ailments I have encountered since I've been treating llamas. My background with llamas began when I was working toward a doctorate in veterinary medicine at Colorado State University. I was introduced to llamas and had some experience treating them under the knowledgeable and enthusiastic instruction of Dr. LaRue Johnson. However, upon graduating in 1983, I chose to enter a primarily equine practice. This practice was located high in the mountains of Colorado, which is ideal "llama country". I had barely begun my practice when a llama owner eagerly requested a veterinarian who might be interested in llamas. Being an energetic new graduate, I accepted her offer.

After the first encounter with her llamas, I became instantly "addicted" to llamas, as I'm sure all of you reading this book are. I was so impressed with their gentle nature and obvious intelligence. I proceeded to follow this admiration of llamas with hours of research, and I attended as many llama veterinary seminars as I could. Gradually, I specialized my practice to include only llamas.

Along the way, I met Ingrid Asmus, the ranch manager for Rocky Mountain Llamas. Ingrid's background included work as a Medical Technologist, training as an Emergency Medical Technician, and experiences teaching outdoor-oriented first aid. She began working for Bobra Goldsmith of Rocky Mountain Llamas in 1981, at first just helping out with routine health care. Gradually, learning by experience, listening, and reading, she developed excellent skills and knowledge in llama care. These skills were put to use in caring for a large herd of llamas. In addition to the reproductive care, medical troubleshooting, and basic training of these llamas, she led llama packing trips in the Colorado Rockies. She always enjoyed drawing and both her knowledge of llamas and artistic skills have been invaluable for this book. Since our experience, and hence the information in this book, are primarily from the Rocky Mountain region, consult with your local veterinarian regarding regional variation in the incidence of llama diseases in your area. (e.g., hyperthermia in warmer climates, liver flukes in warm and moist areas, etc.)

Ingrid and I agreed that a reference health book was greatly needed to lead llama owners in the right direction as health problems occur. This book is a joint effort using each of our areas of expertise in explaining herd health. The disease aspects are written by me, and the non-medical techniques and views are written by Ingrid. The drawings are all by Ingrid, who also helped immensely in the editing process.

No book can substitute for the experience attained by working on and caring for llamas. This book serves as a guide to enable you to begin to observe your llamas more closely and carefully as you continue to learn from fellow llama owners and veterinarians. Since this book is a general reference guide, it is impossible to include every possible injury or illness. Moreover, not all sick llamas "read the book" or have classical symptoms. A close working relationship with your own veterinarian is the most important aspect of health and husbandry. If you are not comfortable doing a procedure by yourself, or are not sure about which parts of this book are applicable to your own situation, get help from your veterinarian. Remember, above all else, "do no harm".

Clare Hoffman

Author's Note:

Due to the increasing popularity of alpacas, some information on alpacas has been added to this Second Edition, First Printing, of *Caring for Llamas and Alpacas*. Since alpacas are so closely related to llamas, most all of the information presented for llamas holds true for alpacas. They share similar anatomy, bodily functions, as well as diseases. In this edition some important similarities between llamas and alpacas are confirmed and alpaca vital statistics are included. A few dissimilarities are also identified. Hopefully, this will help owners in the successful care and management of their alpacas.

NOTES

CHAPTER 1:
BUYING A LLAMA OR ALPACA

With the significant financial investment required to buy a llama, the buyer must make a careful and thorough evaluation of any potential purchase. Otherwise, money may be spent on a llama or alpaca that is not appropriate to one's needs, or, worse yet, on an animal that will be a financial drain due to medical expenses or because it is not able to perform the task for which it was purchased. Such an evaluation has two steps. The first requires that you as the buyer know what you want or need in a llama or alpaca. The second consists of an examination of the animal itself in order to judge whether it can meet the needs you have established.

People use llamas for a wide variety of purposes. Some want a pet, others, a financial investment. Spinners and weavers may want llama or alpaca fiber for their creative work. Some want a llama that can pull a cart or carry children on its back, while others may need a pack animal. Many people are interested in breeding llamas, and more and more people enjoy showing their animals at fairs and stock shows, either in halter or performance classes. What do you want your llama for?

This decision of your llama's purpose obviously goes hand in hand with the choice of its age and sex. Geldings or less expensive males are often ideal pets, fiber producers, or working animals. They can often be very enjoyable to show in performance classes. More expensive stud quality males and breeding females can be used for nearly any purpose, although breeding females are seldom required to do much beyond raising babies or occasionally attending a halter class. Next, you should decide what age llama would best suit your needs both financially and practically. Do you want a youngster, so that you can watch it grow up and then train it? Or, would you prefer to buy an adult that is already a proven breeder or trained and ready to work? Lastly, personal preference about fiber color and length should be considered. Only when you know the answer to these questions are you ready to begin looking for the right llama for you.

Perhaps your interest lies in buying an alpaca. Alpacas are gentle, even-tempered, likeable animals and are easily managed. Alpacas are becoming increasingly popular in the United States and are used for their fiber, breeding, showing and as pets. Alpacas and llamas are very closely related. Both have 74 chromosomes, can interbreed and produce fertile offspring. Alpaca anatomy, physiology and medicine are quite similar to those of their South American cousins, llamas. Alpaca coats come in a variety of colors - from solid to multicolored. Their luxurious fiber is excellent quality, much finer than that of the llama. Alpacas produce 1 to 8 pounds (4 lbs. average) of fiber annually. Of course, diet, disease and genetics can influence the amount and quality of the fiber. There are two breeds of alpacas. The most abundant breed, the huacaya, has crimped (wavy) fiber resulting in a resil-ient, fleecy look. The breed with no crimp in its fiber is called the suri. Suri fiber often hangs in curls.

Visit as many farms as you can, and try to learn as much as possible at each farm. Research thoroughly the prospective llama's background, pedigree, and any offspring produced. Approach any prospects with an open mind. In other words, don't decide to buy any animal no matter what, and avoid impulsive buying just because you can take it home immediately. Your decision to buy a particular llama should be based on many aspects of the animal. You should not buy a llama on the basis of one variable alone. For example, buying a particular llama ONLY because of its pedigree could cause an oversight of a weakness or abnormality in that llama.

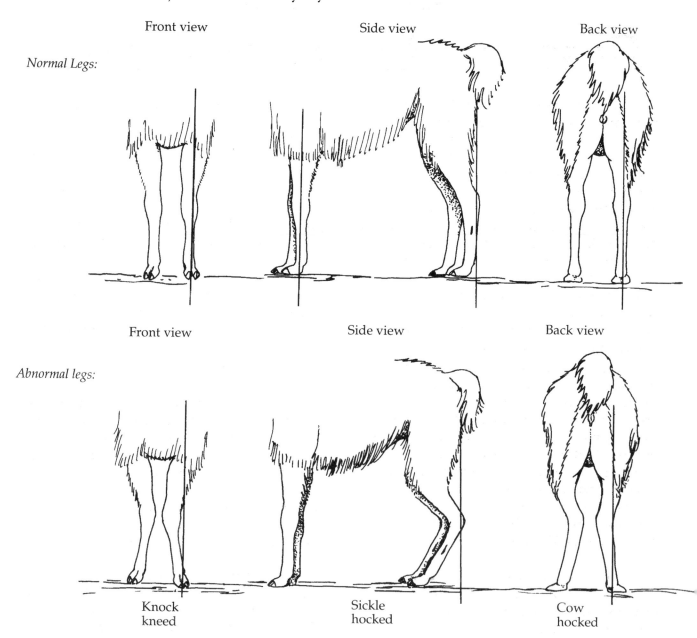

Figure 1.1 **LEG CONFORMATION**
Sometimes it is necessary to shave an animal's legs in order to tell whether they are straight. With experience this becomes less necessary.

14

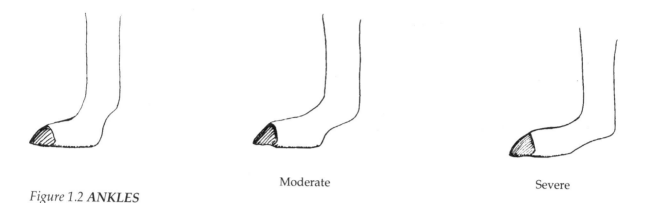

Normal llama foot
(Alpacas have a more verticle ankle)

Abnormal foot, down at the ankle (fetlock) joint due to weakened tendons

Moderate

Severe

Figure 1.2 ANKLES

When you visit a farm to look at the animals that are for sale, first observe the general cleanliness of the environment and the overall condition of the animals. This may provide information about potential problems in nutrition or parasite control. This is not possible when you buy an animal at an auction, but at a well managed sale you should still be able to get background information on any llama or alpaca that interests you. If no information exists, you will be taking a gamble in purchasing any animal.

When you first see an animal that you might want to buy, watch it loose in its own home environment. Compare it to its companions, and watch the way it moves. If the llama is with others of the same age group, compare their size and overall health with the llama you are interested in. Don't ignore a feeling that something just isn't right with an animal, even though you can't pinpoint it. Most of the time you will be correct. Next, ask the seller to catch the animal. This way you can get an idea of the llama's disposition, the degree of training and whether it has any objectionable habits. As the seller is leading the animal towards you, watch for its general attitude or willingness to please. Observe its gait for smoothness and any signs of lameness.

When the llama is nearby, you can observe its legs. Ideally, the legs should be straight. You should be able to draw a straight vertical or plumb line down the front and side of the front legs. A plumb line can also be drawn down the side and back of the rear legs. (Figure 1.1) Also, look at the ankles. The joint should be close to vertical and not drooping towards the ground. (Figure 1.2) Remember, all of this should be viewed in the light of the animal's purpose. Straight legs are more important in breeding or working llamas than they are for strictly fiber producing or pet llamas.

Next, check the skin and coat. Is the fiber a desirable length and color for your needs and interests? Part the fiber and observe the skin. It should be pink or grey, depending on the llama's color. It should not be reddened, bloody, or scaly. Any scaly or flaky skin may indicate poor nutrition, poor general health, or parasitism. Check for lice, and nits (tiny light-colored lice eggs) attached to the fibers. Bald spots should not be present, as these can be an indication of mange, fungal disease, or other skin diseases. As you feel the skin, inquire about any odd lumps or bumps if discovered.

Finally, examine the animal carefully from front to back, checking the following items:

15

- **Head:** The eyes should be clear and not teary. Make sure to note if the animal can see on both sides. It should notice and avoid your hand when held near its face and should not bump into its surroundings. Look into the mouth. The incisors or lower front teeth should be in line with the dental pad on the upper jaw. Your veterinarian will need to examine the mouth more closely, as molars are difficult for an inexperienced person to observe.

  Check for discharge from the nose. It should be dry or only slightly moist. Look at the ears. Are they cosmetically acceptable? Llama ears are erect and should move actively as the llama listens. One ear held abnormally to the side could be indicative of ear problems.

- **Neck:** Check the neck for general balance and centering. Kinks in the neck could indicate old fractures. Look for bald spots and lumps.

- **Body:** Feel the backbone to tell if the llama is obese, thin or well muscled. The back should be straight, not swayed or humped. Look and feel under the belly for an umbilical hernia or small soft sac protruding from the area of the umbilicus on the short-haired underside of the belly.

- **Reproductive structures:** On a female, check the mammary glands or udder. Four teats are normal. Extras may be a minor problem if they don't produce milk and the baby wastes time nursing from a nonfunctional teat. Check the vulva too. It is about 1 to 1-1/2 inches long in an adult llama and about half this length in an adult alpaca, though this may vary depending on the reproductive state. On an intact male, make sure he has two good sized and uniform testicles, as you definitely do not want a breeding male with only one testicle.

- **Legs and feet:** Examine all four legs. Look for recently shaved areas, as these may indicate sites of recent surgery. If you notice some bump or possible defect on the leg or on any part of the body, and are not sure if it is normal, check the same spot on the other side of the body. If the bump is present on the other side and in the same place, it is probably a normal structure.

The foot pads should be examined last. If the llama is not cooperative when you try to examine the feet or you are inexperienced around llamas, you may wish to leave them for your veterinarian to check. If the llama is desensitized on all four feet, you can examine each one. Look for thin and worn areas on the pad. Also, look for splits or fissures, sores and missing or crooked toenails.

If, at the end of this check, the animal seems to be one you might like to own, the next step is to have your veterinarian do a pre-purchase exam. This is crucial when you are preparing to invest thousands of dollars on an animal. No matter how experienced you are with animals, some abnormalities, such as heart murmurs, can only be picked up by a medical expert. The pre-purchase exam should include a check of general conformation and lameness, and if it is for an animal intended for breeding, a reproductive exam should also be performed. A potential stud animal must be checked for two normal testicles. A semen evaluation may be desirable if the male is old enough, especially if there is any doubt about its past reproductive success.

A breeding female should have an internal exam if possible. This includes rectal palpation, if the female is big enough, and providing that proper restraint and the disposition of the llama allows. Ultrasound examination (if available) and determination of pregnancy status are also recommended. If breeding potential is questionable, a more extensive examination may be advisable, and this could include a uterine culture and biopsy. One disadvantage of buying a young female is that she will be too small for an internal examination. Only the external genitalia and vagina can be checked, which means that there is some risk that reproductive problems might be discovered at a later date.

Further tests should include a fecal exam to check for parasites. Tuberculosis and Brucellosis tests are also recommended, and may be required if you will be crossing state lines to bring your new llama home. Depending on where you buy the llama, it may be a good idea to have it tested for Anaplasmosis or Bluetongue. While these tests may not be required for your current purchase, negative tests at this time will be helpful for any future testing required and ensure that your herd will be off to a healthy start. Mange therapy may be needed in some areas. Your veterinarian may also suggest other diagnostic tests such as a complete blood count (CBC) or a serum chemistry if there are indications of some potential problems.

Make sure you have a chance to examine a record of the animal's past medical history, including routine vaccinations and dewormings, reproduction, and any previous medical problems. If you decide to purchase the animal, you should get a copy of this record, because it will help to determine its health management in the future.

Usually, any terms of the sale have already been discussed prior to this point. As negotiations for purchase proceed, a clear understanding must exist between buyer and seller regarding all aspects of the finances and warranties, if any exist. Verbal agreements are often used, but a written legal contract may be advisable when large investments are involved. If you buy a breeding female, determine if there is a warranty as to her reproductive capability. Some breeders guarantee that the female has the necessary anatomy but won't guarantee normal function. If a female is determined to be incapable of reproduction in a preset amount of time, some breeders will replace that female with another one of similar quality or return the money. Other breeders make no such guarantee, and if purchasing a baby that is too young to be examined objectively, a risk must be taken in purchasing that animal. Likewise, any reproductive or performance warranties should be specified between buyer and seller regarding the purchase of a male. The seller may also want a warranty from the buyer as to the intended quality of the llama's care such as the housing, feeding, and the presence of other llamas for companionship. Make sure all aspects of the oral or written contract are agreed upon by both parties.

Finally, based on an evaluation of your exam, the veterinarian's pre-purchase exam, any test results, the animal's background history, and your own needs and interests, you will be ready to make the decision whether or not to purchase the llama or alpaca.

SAMPLE VETERINARY PRE-PURCHASE EXAM FORM

Purchaser _____
Address _____
_____
Phone _____
Seller _____
Address _____
Phone _____
Place of exam _____
Date of exam _____ Time _____
Name of llama _____ Sex _____
Age _____ Sire _____ Dam _____
Approximate weight _____
Tattoo/Brands/Microchip I.D. _____
Color and markings _____
Blood typing _____
Is stabling (environment) adequate? _____
Evidence of bad habits? If so, what? _____
History of disease? _____

PHYSICAL EXAM

1. Temperament and attitude _____
2. Scars or blemishes _____
3. Temperature _____
4. Head
   A. Eyes _____
   B. Ears _____
   C. Lips _____
   D. Tongue _____
   E. Mucous membranes _____
   F. Dentition (missing teeth) _____
   G. Fighting teeth _____
5. Cardiovascular System
   A. Heart auscultation _____
   B. Heart Rate _____
   C. Pulse _____
   D. Capillary refill time _____
   E. Heart auscultation after exercise _____
6. Respiratory system
   A. Nasal discharge _____
   B. Larynx palpation _____
   C. Trachea and lung auscultation _____
   D. Auscultation after exercise _____
7. Digestive system
   A. Ruminations _____
   B. Feces (ova, parasites, consistency, form) ___
8. Skin and fiber
   A. Parasites _____
9. Any hernias present? _____
10. General body symmetry _____
11. Limbs
    A. Left front
       Conformation_____ Pad/toenail _____
       Palpation_____ Lameness _____

B. Right front
   Conformation_____ Pad/toenail_____
   Palpation_____ Lameness _____
C. Left rear
   Conformation_____ Pad/toenail _____
   Palpation_____ Lameness _____
D. Right rear
   Conformation_____ Pad/toenail _____
   Palpation_____ Lameness _____
12. Reproductive Structures
    A. Male
       1. Castrated?_____
       2. History of impregnating females?_____
       3. Testicles _____
       4. Penis _____
       5. Sheath _____
    B. Female
       1. External genitalia _____
       2. Mammary glands _____
       3. History of previous pregnancies?_____
       4. Rectal palpation _____
       5. Ultrasound exam _____
       6. Vaginal exam (digital, vaginoscopic)_____
       7. Pregnant?_____
       8. Other (culture, biopsy)_____
13. Ancillary tests
    A. Progesterone assay (pregnancy)_____
    B. Skin scrapings _____
    C. Cultures _____
    D. Urinalysis _____
    E. CBC _____
    F. Serum chemistries _____
    G. T.B. test _____
    H. Brucellosis _____
    1. Anaplasmosis _____
    J. Bluetongue _____
    K. Other _____
14. Preventative medicine history
    A. Vaccinations _____
    B. Dewormings _____
    C. External parasite control _____
    D. Other _____
General comments and observations _____
_____
_____
_____
_____

_____
Signature

CHAPTER 2:
CONCERNS OF TRAVELING

Introduction

Llamas and alpacas are spending more time on the road these days, traveling to or from shows or sales, the veterinarian, breeding engagements or the trailhead. With all this time on the road, it is important to recognize that travel is stressful and potentially dangerous for the animals. The strange environment, different companions or no companions, strange arrangements for food and water, temperature extremes, noise and movement all contribute to the problem. There are a number of things that can be done to limit both stress and hazards, and so contribute to the well-being of the animals being transported.

Official Health Certificate

With shows and sales becoming more and more popular, llamas and alpacas are spending more time on the road. There are certain considerations for safe and legal travel. Most states require official health certificates to be completed before allowing entry into their state. Check with your veterinarian as to the specific requirements for your state of destination. These certificates must be completed by an accredited veterinarian following a physical examination of the llama. Many states also require tests for Tuberculosis, Brucellosis, Bluetongue, or Anaplasmosis. Some states may require mange treatment or other special proce-

dures. Often, the tests take several days to complete. After all tests are complete, some states require a permit number to appear on the health certificate, which may require additional time.

DO NOT wait until the last minute to get your tests and health certificates for a trip. In fact, it is best to notify your veterinarian well in advance, about four weeks before you plan to leave. In general, a health certificate, once issued, is good for thirty days. This varies from one state to the next. You must meet the laws of the state of your destination, not the states that you pass through on the way.

Insurance

Insurance is not required for transporting llamas or alpacas. However, as the distance and duration of the trip increase, and depending on the number and value of the animals being carried, insurance becomes more and more important. Insurance is available to cover brief periods including transportation and settling in to a new environment[1], so check with an animal insurer.

Concerns of Traveling

Llamas and alpacas are very adaptable animals. They have been successfully transported for varying distances in station wagons, jeeps, pickup trucks, horse tailers, boats and even small airplanes. In general, the most common and convenient methods of transportation are trailers, trucks and vans.

Each type of vehicle has its own advantages and disadvantages. The floors of many vans and pickup trucks are high off the ground. While most llamas and alpacas can learn to jump right in, it may be helpful to have a loading platform or short loading board to ease the transition. Trailers are generally set low, and both kinds of animals can learn to step directly in.

No matter what type of vehicle you use, it is important to have a non-slippery floor. Most trailers have a wooden floor which works fine. Pickups usually have metal floors that are slippery when wet, and may become uncomfortably hot or cold. Bedding can be used to add comfort for the llama. Straw can be used, but in an open vehicle, you are likely to be trailing blades of straw for miles. An alternative is rubber-backed, indoor/outdoor carpeting. Several types of rubber mats work well, too. However, black rubber can become very hot in the back of an open pickup on a hot day.

Untrained animals can be lifted or boosted into a trailer or other vehicle, given enough people to help. However, training animals to load is not difficult, and makes transport much easier. It also makes travel less stressful, and provides more options in case of emergencies. Short practice trips get your llama or alpaca used to the idea of travel, and allow you to learn more about how a particular animal will respond. Many llamas and alpacas will travel lying down in kush position. Some will stand up the whole way, while others will be restless. Even a group of untrained animals leaving home for the first time will usually settle down once the vehicle begins to move, snuggling together for moral support. However, there may be quite a bit of moving around at first. It is useful to be able to observe your animals easily, to be aware of habitual travel patterns, and of problems as they develop.

It is not a good idea to tie llamas or alpacas during travel. A number of llama and alpaca deaths have been reported as a direct result of being tied in trailers. The only time you should consider breaking this rule is if you are making a short trip with well trained, experienced animals that are accustomed to being tied. In this case, tying may help to minimize encounters with other llamas, or keep the animal's back to the wind. A mother llama or alpaca could be tied, if she is trained, while her infant is left loose. If an animal must be tied, the slack in the rope should be just enough to allow it to stand up or lie down. If the rope is too loose, or the animal is not used to being tied, it is all too easy for it to get the rope twisted around its neck. At best, this is uncomfortable; at worst, fatal. Never carry a mixed load, where some animals are tied and others left free. This can result in serious tangles, fights and, possibly, injuries.

If a group of untrained mature animals who are strangers to each other must be transported together, the best choice would be to physically separate them with some sort of divider. Well secured chain link or plywood panels work well for this purpose. When used for llamas or alpacas, dividers should be floor to ceiling, or have only a narrow gap below the ceiling. A divider that ends several feet above the floor, such as those in some horse trailers, may cause problems if a llama or alpaca lying down in kush position underneath it tries to stand up suddenly. Fights could occur under high dividers, or over low ones. Since dividers are designed to prevent problems, make sure that what you use will work efficiently to separate animals and not cause more problems.

Another possible danger in some trailers, especially some types of horse trailers, is the large opening that exists over the doors at the back. A frightened, young or inexperienced llama or alpaca could try to jump out, and might succeed. Such openings should be covered with wire netting or some other visible and substantial barrier.

Animals being transported across the country can, if necessary, be left in a trailer or truck for several days. However, they must be fed and watered at least each morning and evening, and temperatures in the vehicle must be monitored and kept within a reasonable range. Leaving llamas and alpacas in a vehicle of several days may be unavoidable if a single person must transport a group of animals that are not trained to lead or load, but travel of this kind will become more and more stressful for everyone involved, human or animal, the longer the trip.

When carrying mothers and nursing babies, it may be essential to stop every couple of hours. Some mothers will stay in kush position, preventing their babies from nursing, as long as the vehicle is moving.

As anyone knows who has traveled across country, remaining in the vehicle can be boring and uncomfortable. Therefore, if the situation allows, stop briefly every few hours for a short walk and a "rest stop" at an appropriate spot. If you carry a coffee can with fresh fecal pellets in it, you can scatter the pellets on the ground in a selected place to show the llama or alpaca that this is a good place to relieve itself. Don't forget to gather a few more pellets for the next stop! Such rest stops will keep the animal more comfortable, and your cleanup will be easier.

Specific Types of Vehicles

- **Horse trailers:** Horse trailers are readily available, either used or new. If you buy a used trailer, no matter what kind, be SURE to check the floor, as it is often the most worn part of the trailer. Continue to check the floor regularly, or before each use. Horse trailers are sturdy and can be easily used to transport llamas and alpacas. Without a divider, three or even four llamas can be carried in a two horse trailer. Horse trailers already have ventilation built in, and are sturdy and quite safe. On the negative side, horse trailers are expensive and quite heavy, requiring a powerful vehicle to pull them. Unless you already own one, or wish to have one so that you can carry both horses and llamas, a horse trailer may be more trailer than you really need.

- **Other trailers:** Utility trailers, such as the kind used by U-Haul® and other rental companies, may occasionally be found for sale. They MUST be extensively modified to provide safe llama or alpaca transport. Ventilation is absolutely essential, and may be combined with the addition of windows,

since llamas and alpacas enjoy watching the world go by as they travel. Windows also make it possible to monitor the behavior and well-being of the animals without constantly opening the trailer. Insulation is also vital, especially in metal trailers, lest the trailer turn into an oven or an ice-box. Make sure there are no exposed sharp edges or corners, and that there are enough places to tie animals temporarily while loading. Addition of interior lights may make it easier to use the trailer at night, or to cope with emergencies.

With these modifications, such trailers may become an excellent means of carrying llamas and alpacas. The trailers are not as heavy or sturdy as horse trailers, but the sizes work very well for llamas or alpacas. A five-by-eight foot trailer with a five foot, six inch ceiling can be used for two to four llamas, while a six-by-twelve foot trailer can carry up to seven llamas.

- **Specialized llama or alpaca trailers:** In recent years, a number of trailers have been developed especially for transporting llamas or alpacas. These may provide all the features described above and could even include air conditioning and other conveniences. If you anticipate traveling extensively with your animals, these may be well worth investigating.

- **Pickup trucks with stock racks:** Putting a stock rack on a pickup truck is often the fastest and cheapest way to create a transportation vehicle for a llama or alpaca. Even a small pickup may be adequate, with the addition of a stock rack and non-skid mat, to transport two adult llamas or several alpacas or youngsters. Stock racks may be purchased new or used, or fabricated from metal or wood. It is important that the openings in the rack be narrow, so that a llama or alpaca won't try to escape, or get its leg caught. Some animals, particularly youngsters, may try to jump over the top if it is left uncovered. If the truck will be used in bad weather or for long distance traveling, the stock rack should be covered. Pickups with stock racks are well ventilated.

In fact, problems are more likely to occur because of wind irritation to the animal's eyes. The wind also blows feed and bedding around. These problems mean that the pickup with a stock rack is probably the least suitable vehicle for long distance transportation of llamas and alpacas. Such vehicles are better for short distance trips in mild weather.

- **Vans:** Vans can make transportation of llamas and alpacas very easy. With the back seats removed, the animals can ride in the back, the people in front, making the trip a real family affair. An extra piece of carpet can protect the interior, as well as insulate the animals from the heat of the transmission during hot weather. Frequent rest stops with a pellet can makes clean up minimal in an easy access van. A chain link partition between animals and people is probably desirable, since it will keep the llama or alpaca from stealing your lunch, and prevent the abrupt arrival of an animal in the front seat during a sudden stop. If your van is air conditioned, it can provide a good way to prevent heat stress in hot weather. On the negative side, if you are transporting two animals who don't enjoy each other's company, they may spit at each other, and then you too will share the odor.

Health Care on the Road

Virtually every health care problem described in this book may occur on the road. Some may be much more likely to take place while traveling. These include injuries due to loading or unloading, inappropriate use of ropes, falls in the vehicle, problems with heat and cold, digestive problems due to irregular or unaccustomed feed and water or unfamiliar toxic plants, choking, and respiratory problems caused by dust or inadequate ventilation. Close exposure to other animals and the stress of travel may make transmission of illness more likely. For all these reasons, it is important to monitor animals frequently, and be prepared to act. Have a first aid kit available. Also bring along emergency phone numbers, so that you can find help wherever the problem may occur.

Travel with llamas and alpacas is usually enjoyable and rewarding for all. It is not particularly difficult or demanding, but like any complex activity, does require common sense. Given the opportunity to consider the kinds of problems or emergencies that may occur due to injury, illness, mechanical accident or bad weather, you can take steps to prevent, or deal with, whatever may come up.

CHAPTER 3:
RESTRAINT

Introduction

Whenever medical work is done on a llama or alpaca, restraint will usually be necessary, as it allows necessary procedures to be done efficiently without the animal or the people getting hurt. The kind of restraint needed will vary to some extent from one animal to the next, and between species. Most llamas are trained to some extent, and are usually fairly easy to work with. Alpacas are equally easy to train, but since they are used primarily for fiber and breeding purposes, they are often left untrained. For untrained animals, management and care techniques will be different, and depend less on voluntary cooperation of the animal, and more on control of its options and movements by human handlers. The restraint required will also depend on the type of treatment to be given. For example, a llama who stands quietly to be dosed with a deworming paste may feel much less cooperative when the time comes for rectal palpation! Because of this, it is useful to have a variety of options available as circumstances demand. In general, don't use more restraint than you need to accomplish a given task. Also, don't keep the animal restrained any longer than you need to. Finally, remember that steady, quiet handling and reassurance should be a part of any system of restraint.

Restraint without Equipment

From time to time it will be necessary to vaccinate or medicate a young llama or alpaca that has not been halter broken. For newborns and infants, the straddle position (illustrated in Chapter 20, without any weight resting on the infant), may be used for intubation or other procedures. Alternatively, if someone can still pick up and hold the youngster, it is often easiest to do just that. One person can hold the baby off the ground, one arm around the chest, and the other around the rump. For a larger animal, one arm can go around the chest while the other goes under the belly, in front of the hind legs. This leaves enough space for a second person to inject or otherwise treat the llama or alpaca.

If the young llama or alpaca is too large to lift, an alternative method is for two or even three people to push it against a wall. It is important to use a wall, not a fence, since it limits the number of places the animal can use as footholds in its effort to get away, making your task easier and safer. Often, especially with smaller animals, it is possible for each of the people to put a knee up under the animal's chest or belly, and actually lift it slightly off the ground. In this position, it will be fairly easy for someone to deworm or vaccinate the animal, as appropriate.

Depending on its size and temperament, a mature but untrained alpaca may be re-

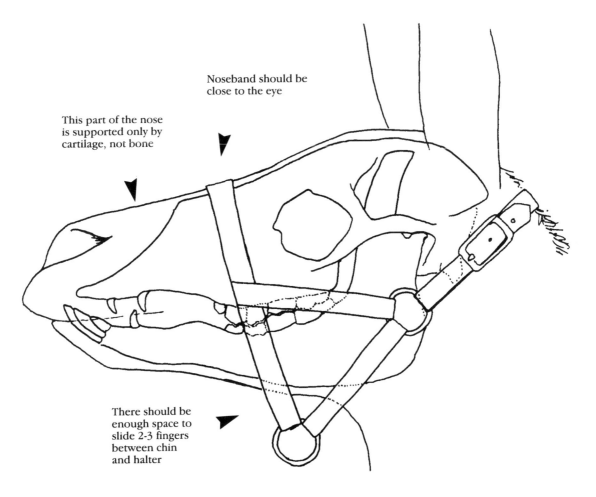

Noseband should be
close to the eye

This part of the nose
is supported only by
cartilage, not bone

There should be
enough space to
slide 2-3 fingers
between chin
and halter

*Figure 3.1 **PROPER FIT OF HALTER***

strained by holding its head against the handler's body with one hand, while using the other to hold either the withers or tail.[1] More complex procedures will require further restraint.

Halters

It is possible to put a halter on a completely untrained llama or alpaca solely in order to treat or vaccinate it, but the results are seldom worth the effort. The animal may be so upset by the halter, lead rope, and being tied or held in close proximity to a human, that it may panic, risking injury to itself and any handlers. It is better to go through training in a measured, reasonable fashion and give routine medical care when the llama or alpaca accepts the halter and lead rope as parts of everyday life.

Ideally, any llama or alpaca past weaning age should be halter-broken and trained to lead. When a halter is used, it must be adjusted properly in order to be both safe and effective. No matter what kind of halter it is, the noseband must be placed high on the nose, close to the eye. Only in this position will the halter be resting on bone. Much of the llama's and alpaca's nose is cartilage and soft tissue. A halter that slides down or is placed on cartilage can easily compress the nose, just as a fingertip pushing down on the tip of your nose can compress and deform it. This is uncomfortable for the llama and dangerous, because it can block or partially block the airway. Since these animals normally breathe only through the nose, obstruction of the nasal passages causes panic. No animal can be expected to stand quietly or safely while being suffocated by its halter.

24

It is important to make sure that the halter fits correctly from the beginning, and if the halter is adjustable, to keep checking it periodically if the procedure is long or the llama is unruly. It is also a good idea to avoid bovine or heifer-type halters which cross under the chin. When there is a strong pull on the lead rope, these can tighten around the nose, starting a vicious cycle. As the halter becomes tighter and more painful, the llama's struggle increases.

Halters must also be loose enough that the llama or alpaca can still open its mouth to eat or chew with the halter on. It should be easy to slide two or three fingers between the halter and the animal's lower jaw. A halter that is much looser than this may be too loose to be functional. Halters should be used only when needed, not just put on and left on because it is easier. Show halters may fit more snugly for very brief periods, but if an animal must, for some specific reason, wear its halter for days at a time, it may be more comfortable with a slightly looser halter.

Knots

Once the llama is content wearing a halter, has been taught to lead, and accept the restraint of the lead rope, the next step is to tie it to something. This is often a challenge to some people whose main experience with knots is tying their shoes. However, even though there are hundreds of different knots for different purposes, it is seldom really necessary to know more than a few. It is important to know not only how to tie the knot, but when it is appropriate to use it. For example, a square knot is excellent for tying a cinch to a fence rail in order to keep a llama in a chute from lying down. If, however, it is used to tie a llama to a tree where all the tension on the knot comes from one end of the rope, it will turn into a slip knot. For this purpose, a quick release knot or bowline are superior. It is a good idea, whenever using ropes for restraint of any kind, to keep a sharp pocket knife handy. If, for whatever reason, a llama is endangered by a rope, the rope can be cut without any wasted time. Five of the most useful knots are illustrated and described below.

Square Knot

The bow tying a shoe is related very closely to a square knot, and this shows one of its most popular characteristics. Properly tied, it will not slip under strain, unlike its cousin, the granny knot.

Figure 3.2 SQUARE KNOT

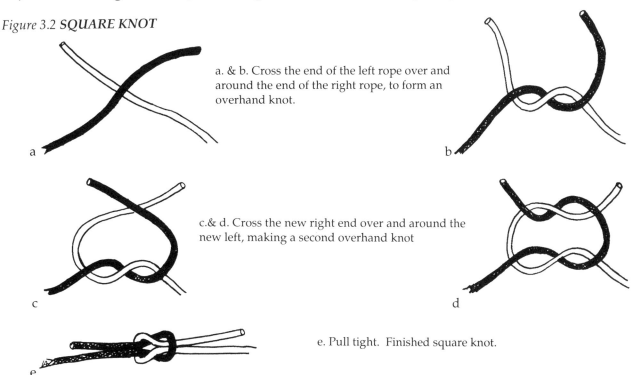

a. & b. Cross the end of the left rope over and around the end of the right rope, to form an overhand knot.

c.& d. Cross the new right end over and around the new left, making a second overhand knot

e. Pull tight. Finished square knot.

Bowline

This strong and reliable knot has been used for generations by sailors and climbers in situations where lives may depend on the knot staying tied and not slipping until it is untied, no matter how much strain is put on it. It is often used to tie a stake line to a tree or stake pin, so that a llama can graze safely for hours or overnight. It can always be untied easily.

Figure 3.3 **BOWLINE**

a. Wrap rope around fencepost or through a ring. Form a loop in the standing part of the rope.

b. Pass the free end of the rope underneath and through the loop and around standing part of rope.

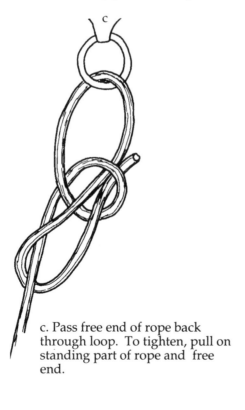

c. Pass free end of rope back through loop. To tighten, pull on standing part of rope and free end.

d. Finished bowline knot.

Quick Release

Often used when a llama will be tied for a brief time, the quick release knot, as its name implies, can be untied quickly. It will remain stable, but the knot will get tighter if the llama pulls on it, and it is not as reliable for long term use as the bowline.

Clove Hitch

This knot is best used to secure a rope with a fast release snap attached to the vertical posts of a chute for cross tying, or to a stanchion. If the rope tightens and relaxes frequently, the knot will become loose and unreliable.

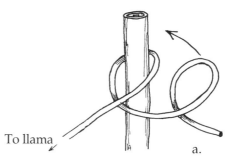

a. Form loop around post.

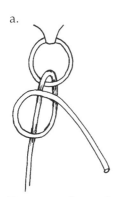

a. Pass rope through ring or around a post. Make a loop in the free end of the rope, and place it over the standing part.

b. Pass the free end behind both the standing part of the rope and the loop. Push a bight or loop through the first loop.

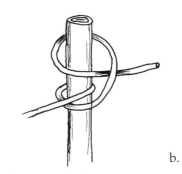

b. Form second loop and pass it over top of post. Pull both ends of rope to tighten.

c. Pull on loop to tighten the knot.

d. To keep the animal from pulling on the free end of the rope and untying the knot, pass the end through the loop.

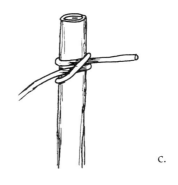

Figure 3.4 QUICK RELEASE KNOT

Figure 3.5 CLOVE HITCH

Lark's Head

This simple knot is often used to attach a cinch to a saddle or rope.

a.

a. Pass bight of rope through ring to form a loop.

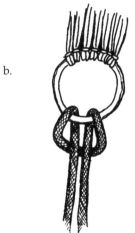

b.

b. Pass free ends of rope through loop, and pull to tighten.

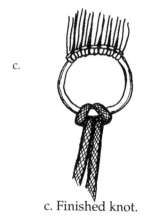

c.

c. Finished knot.

Figure 3.6 **LARK'S HEAD**

Cross Tying

One of the simplest ways to restrain an animal is to cross tie it. Trees, fence posts or whatever is available may be used. When an animal is cross-tied, it will not be able to move forwards or backwards very far. While it can still swing its hind quarters from side to side, its head will be fairly stable. This may be the only restraint needed to deworm an animal, treat an eye injury, or for other simple procedures. Cross-tying may also be used in conjunction with other, more sophisticated methods of restraint, such as chutes.

Figure 3.7 **CROSS-TIE**

Chutes

Chutes are narrow passages or enclosures in which an animal can be safely immobilized for a brief period to allow speedy and efficient care. A number of different kinds of chutes are available on the market, and others can be constructed inexpensively and easily. (Instructions for two of these may be found in Appendix IV.) Each chute has its own advantages and disadvantages. Whether a particular chute is right for you will depend in part on whether you work with llamas or alpacas, and on the level of training of the animals. It will depend too on the way you intend to use the chute. Will it be used exclusively for medical care, or would you like to use it for grooming or shearing as well? For medical care, a chute which allows access only to the animal's head, neck and rump may be sufficient, but for purposes of injections, milking, and nail trimming as well as grooming or shearing, a chute which provides access to the animal's sides is very useful.

If you are planning to buy or construct a chute, talk to people who work with the animal that interests you, and who have experience with various kinds of chutes. In this way, you can find out what would be most useful to you, and avoid or take steps to correct any potential problems.

No matter what kind of chute you eventually use, there are a few basic precautions to take. Make sure that any sharp edges or corners are eliminated or covered and padded. These seldom cause injury during quiet routine work, but during a brief, intense struggle, you or the animal may have very little choice about where some sensitive part of the anatomy ends up.

More important, however, is the need to make sure that the chute is stable and will not tip over even if a terrified animal attempts to leap out and throws all its weight on one side. If a chute falls over, there is a high risk of injury to the animal inside, and to anyone attempting to treat the animal. Follow the manufacturer's or designer's instructions to make sure your chute cannot be knocked over.

If a chute appears to be unstable, don't purchase or use it.

Remember that it is usually best to use the minimum restraint necessary, both in terms of the amount of time required to set up the restraint system, and in terms of the animal's comfort and the risk of injury caused by the restraint system itself. It is rarely necessary to completely immobilize a llama or alpaca. Often it is best to move with the animal and be flexible, rather than wasting effort on trying to keep it totally motionless. Many, if not most, llamas or alpacas can be vaccinated, dewormed, or even have their fighting teeth cut while simply being cross-tied in a chute.

Sometimes a llama or alpaca will be so offended by even the possibility of being injected, milked, or whatever, that it will drop into the kush position. If the animal is comfortable in this position, and you can do so, it is fine to continue the procedure. However, if the animal is hanging itself from the cross-tie, you must get it up quickly, or unfasten the cross-tie. A panic snap on the cross-tie may be very helpful at this point, but it may be better to prevent the animal from lying down in the first place. This is easily done in some chutes which have cinches built into the design, but any open-sided chute can be adapted to do the same task. In order to do this, you will need two saddle cinches, a short length of cord, and four spare lead ropes or ropes about ten feet long. Use the cord to tie the cinches eight to ten inches apart (less for alpacas), as this will prevent the back cinch from sliding too far back and irritating the penis sheath or udder. Each rope is tied to one cinch ring with a lark's head knot. The ropes on one side of the animal will be tied to the top rail of the chute, positioned so that the cinches will be centered under the animal's belly, with the front cinch directly behind the front legs, and the back one about halfway back on the stomach. The remaining ropes are passed under the animal's belly, and up over the top rail on the near side of the chute, pulling the cinches behind them. When the cinches are snug under the llama, the ropes should be tied firmly with quick release knots or square

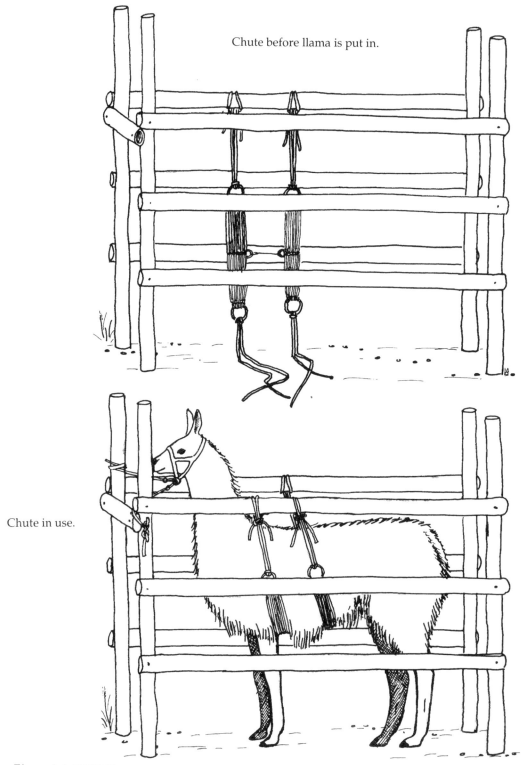

Chute before llama is put in.

Chute in use.

Figure 3.8 **CHUTE**

30

knots. Do not try to lift the animal off the ground by over-tightening the cinches. An animal trying to lie down will instead be cradled by the cinches. Don't leave an animal in this position unattended, or for more than a few minutes, as severe internal injury can result. Use caution with cinches that are attached to the chute by chains, because they can be almost impossible to adjust or unfasten under tension, and unlike ropes, they can-

just behind the neck, and tied to the lower rails of the chute. For a really inventive animal, who may try any and all of these methods, often in rapid succession, the single cinch over the shoulders may be combined with the double cinches described previously.

Finally, for mature animals who must be intubated, or have a blood sample drawn, or the occasional male who strongly resists tooth cutting, a stanchion should be used. In order

Figure 3.9 **CHUTE**

not be cut in an emergency. Keeping the animal up in this way may be particularly useful if that is the position your veterinarian prefers for rectal palpation. It may also be valuable when a llama or alpaca uses passive resistance to avoid necessary care.

By contrast, other llamas or alpacas will try to escape by rearing or jumping. Tying the animal's head low to the stanchion or the front beam of the chute may prevent this. If not, a single cinch with ropes tied as described above can be positioned over the animal's shoulders,

to be able to use this method of restraint, it is necessary to have a chute equipped with two poles or posts at the head end. One or both of these must pivot at ground level. It is essential that the moveable posts can be fixed firmly in place with a pin or chain once the llama or alpaca is in place. These posts are not intended to hold the animal's neck in a tight pinch. Instead, they are meant to prevent forward movement by pressing against the shoulder, and need only be close enough to prevent the animal's squeezing between them. Make sure, when working with a stanchion chute, that the posts are far enough apart that the animal's neck will not be pinched, even if it drops into kush position.

The animal is led forward into the stanchion chute, and its lead rope is passed between the two moveable posts. While one person maintains tension on the lead rope, the other must bring the moveable posts in close to the animal's neck, so that its neck is sandwiched between the two posts. The moveable parts are then fixed in position. During this time it is especially important that forward tension be maintained on the animal's lead rope, as otherwise it may attempt to back out of the chute, risking damage to the sides of its head and eyes on the way. Once the neck is stabilized between the two posts, the animal's head is pulled forward and tied securely, preferably to two posts rather than just one. These ties prevent the animal from moving backwards, and complete the restraint system provided by the stanchion. A panic snap is a good idea on one or both of the ropes.

When the medical care is completed, it is equally important to release the animal with care. The moveable posts can be loosened and pushed to the side. This allows the llama to move forward and out of the chute. Alternatively, with the posts moved out of the way, the ropes can be untied, and the llama or alpaca allowed to back out of the chute.

An important component of any method or restraint is the use of verbal and nonverbal commands and reassurance. Movements should be calm, quiet and steady as opposed to tense and jerky. Conversation or commands should be quiet and definite, not noisy and confused. Working in this way will reassure the animal, and avoid upsetting it more than necessary. Sometimes having someone at the animal's head to cuddle it, rub it behind the ears, and talk sweetly to it will be enough to calm it so that a procedure can be completed. Naturally, things do go wrong sometimes, but often the best course will be to release the animal briefly, let it settle down, and then try again, rather than trying to force it to submit then and there.

Don't use a method of restraint that is more restrictive than necessary for the particular animal and procedure you have in mind. Avoid using even necessary restraints for any longer than they are actually needed. Injuries and deaths have been caused by leaving llamas and alpacas in tight restraint for too long. Use the restraint you need to give the necessary care, and then let the animal go.

Figure 3.10 LLAMA IN STANCHION

CHAPTER 4:

NUTRITION

Introduction

A separate book could be written just about nutrition. This chapter is only a brief introduction. Adult llamas and alpacas are modified ruminants, which means that they regurgitate food and chew cud but have a somewhat different stomach anatomy than conventional ruminants. Llamas and alpacas are very efficient at utilizing available feed, making them relatively inexpensive large animals to maintain.

Water

Water is the most important nutrient for your llama. At home, provide adequate good quality clean water at all times. When out on the trail, pause at water crossings to allow your llama to drink if he wants. (Don't be alarmed if your llama doesn't drink when you think he should be thirsty.) If you are packing in an arid area, make sure he gets at LEAST one big drink of clean water each day. The average healthy adult llama drinks 3/4 to 3-1/2 gallons of water each day. The average adult alpaca drinks 1 to 1-1/2 quarts of water each day.[1] Water needs will increase with exercise, lactation, and high environmental temperature.

Decreased water intake causes decreased food intake which leads to a lowered overall performance of the animal, including decreased growth rate and milk production. Also, less water is eliminated in the feces, resulting in dry feces that are more likely to cause an impaction (see Chapter 14). Decreased water intake may be caused by water that is too warm, too cold, frozen, inaccessible, distasteful, or poor in quality. Your water should be evaluated for harmful amounts of contaminants such as nitrates, lead, selenium, arsenic, and mercury to determine its suitability for llamas. In addition, water contaminated with microorganisms should be avoided. For example, stagnant pond water is often an excellent source for *Leptospira* contamination (see Chapter 5). Water with heavy algae growth is harmful as some blue-green algae may result in poisoning.

Energy

Energy-providing feeds are the next most important in nutritive value. Energy is available from carbohydrates, fats, and proteins. It is found in feeds such as pasture forage, hay, grains, and pelleted feeds. To ensure good quality feed, avoid dusty, moldy, foul-smelling, and yellow hay. If you rely on pasture forage, have it analyzed for its quality and nutritive value (See Appendix VIII.) If you feed a completely formulated and prepared ration, make sure it contains enough roughage (at least 25%), if no other roughage source is provided.[2] Guidelines for feed consumption vary with llama's age, work, weather, gestation, and lactation. Dry matter feed (feed dried to a constant weight) intake in the llama is 1 to 2% of its body weight per day.[3,4] Thus, a 300-pound llama can eat 3 to 6 pounds of dry matter per day. Similarly, alpacas need 1-1/4 to 1-1/2 percent of their body weight in dry matter intake.[5] A 150-pound alpaca can eat about 2 pounds of dry matter per day.

Protein

Protein needs depend on the llama's age, exercise, pregnancy status, and lactation. Protein requirements in llamas and alpacas are fairly low due to ruminants' ability to utilize protein by-products (urea and ammonia). Adult llamas on a maintenance diet only need 8 to 10% protein in the diet. Pregnant and lactating llamas may need 12 to 14% protein. Babies and growing llamas may need from 10 to 16% protein.[6] The protein content of your pasture, hay, or grain should be analyzed by a feed laboratory (See Appendix VIII). As a basis for comparison, grass hay is usually 7 to 10% protein. Alfalfa hay is about 16% protein. The percentage depends on the cutting of the hay, with the first cutting having the lowest protein content. Cereal grains (corn, oats) have a protein content of about 8 to 10%.

Excessive feeding of protein is not recommended. If you overfeed, the excess will be converted to energy, and if the energy is not needed, it is stored as fat. Using protein instead of carbohydrates for energy also seems to contribute to heat stress and should be avoided in hot environments.

Minerals

Minerals of major concern in the animal are salt, calcium, phosphorus, and selenium. To ensure adequate salt intake, loose salt should be available free choice. It should be placed out of the weather to avoid wetness.

Calcium, the most abundant mineral in the body, is needed for normal development and maintenance of bones and teeth, for milk production and for other important body functions such as blood clotting. Phosphorus is important in bone structure too, and is an essential component of chemical energy in the body. A calcium-to-phosphorus ratio in the llama's diet should range from 1:1 to 2:1. Calcium and phosphorus needs increase with pregnancy, lactation and growth. Calcium and phosphorus may need to be supplemented in certain cases. Imbalances may prevent normal bone growth, decrease milk production, contribute to crooked legs, osteoporosis, urinary stones and rickets.

Rickets is a condition of very young and growing animals whereby mineralization of bone is abnormal. It is caused by Vitamin D and/or phosphorus deficiency, but the reasons for the deficiency vary. An obvious reason is poor diet. Less evident is rickets in youngsters with adequate diets. These animals likely have abnormal absorption or metabolism of vitamin D and/or phosphorus or have insufficient exposure to sunlight. Signs of rickets include lameness, bone deformities, poor growth rates, enlarged joints and swellings at the lower ends of the ribs. These signs have been observed in alpaca crias and may be accompanied by an abnormal curved-back stance.[7] Further support of a diagnosis of rickets is obtained through blood tests (phosphorous and calcium levels) and radiographs (X-rays). Your veterinarian will determine which nutritional supplements are necessary.

In general, cereal grains are a good source of phosphorus. Legumes such as clover and alfalfa, or fresh green grass are good sources of calcium. Your present feed ration should be analyzed to determine if it is imbalanced. If calcium needs to be added to your ration (e.g., you feed mostly grass hay), 50/50 dicalcium phosphate and salt can be fed free choice. If phosphorus needs to be added to your ration (e.g., you feed mostly alfalfa hay), cereal grain can be added. If you cannot feed grain because the llama is too fat, the phosphorus may also be supplemented by a phosphorus mineral block made available free choice, or a free choice 50/50 trace mineralized salt and bonemeal supplement, or a free choice 50/50 trace mineralized salt and monosodium phosphate supplement.[8]

34

Another important mineral to consider in your feed program is selenium. Selenium is important for normal immunity, reproductive health, tissue healing and protection against harmful chemicals. Selenium supplementation is ONLY necessary in selenium deficient areas. Check with your local veterinarian or extension agent to determine the selenium content in the soil in your area. You may discover that the soil is dangerously high or deficient in selenium.

If you are in a selenium deficient area, 1 to 1-1/2 mg of selenium per day as an oral supplement is sufficient per adult llama.[9] Selenium is also available in an injectable form if you prefer. Selenium deficiency can cause white muscle disease. With this disease, there is a degeneration of the muscles, including the heart muscle. Signs include muscle stiffness, leg weakness, uncoordination, and difficulty in rising. Exercise will worsen the signs. Due to tongue muscle involvement, affected babies may be unable to nurse. White muscle disease can cause sudden death or lead to death if not treated early enough.

Nutritional zinc requirements are unknown in llamas at this time. Zinc is needed for numerous body functions including normal development, growth, and healing. The normal zinc levels in the llama's body are also unknown, making recognition of zinc deficiency very difficult. Based on signs of zinc deficiency in other species, it should be suspected in llamas with prolonged skin problems (see Chapter 10), poor appetite, decreased growth rates, or reproductive imparities.[10]

Likewise, normal values for copper are not known in llamas. Copper is needed for normal bone, hair and red blood cell production. Copper deficiencies are seen in other species, especially in hot and humid regions, and are usually accompanied by excessive levels of molybdenum in the body. Signs of copper deficiency in other species include anemia, decreased growth rate, and abnormal hair coat.[11]

Vitamins

Vitamins are needed in small amounts for basic body function or metabolism. Some are made in the body while others are needed in the diet. Deficiencies can result in poor growth, poor reproductive abilities, and a depressed immune system. B vitamins are produced by bacteria normally living in the digestive tract of ruminants. Vitamin C is likely produced by the llama's liver, as it is in most animals (primates excluded). These vitamins need not be supplemented in a healthy llama. Vitamin K is probably produced in sufficient quantities in the intestine, but a small amount is still recommended in the diet. If a good quality green hay or pasture forage is fed, the llama can get adequate amounts of vitamins A, D, E, and K.

The major vitamin of concern in the llama is vitamin A. Beta-carotene, a precursor to vitamin A, which is used by the body to produce vitamin A, is present in green forage. The vitamin A content decreases as the forage matures or is stored for long periods of time. The amount of vitamin A precursor can be judged by the green color of the forage. If the forage is green, the vitamin A content is sufficient, if the forage is yellow or brown, there is not enough vitamin A. Vitamin A deficiencies are remedied by injections, oral supplementation, or a change to a better feed. Adequate vitamin A is especially important during late pregnancy, lactation, and rapid growth periods. A newborn has very low levels of vitamin A, but colostrum is rich in vitamin A, and milk is rich in vitamins D and E.

Guidelines for Feeding Your Llama or Alpaca

Whenever changing types or amounts of feeds in ANY animal, do it GRADUALLY. Abrupt feed changes can cause diarrhea and bellyaches.

1. **The Mature Animal:** The average 300 pound healthy llama needs 3 to 6 pounds

of roughage (hay, pasture) per day for maintenance. A 150-pound alpaca can eat about 2 pounds of dry matter per day. Amounts vary depending on the quality of roughage and condition of the llama. As more energy is needed for physical activity or for keeping warm, grain can be added which will replace roughage intake. The amount of grain varies with the individual and its body condition.

2. **The Pregnant Llama:** During the first two-thirds of pregnancy, the llama should be fed the same ration as a mature llama on a maintenance diet. Use care in not overfeeding your llama during this period, as a fat llama tends to have decreased milk production and an increased chance of dystocia (difficult birth). Most of the fetal growth occurs during the last 3 months of pregnancy. All nutrient amounts should be increased proportionally during these three months. If good quality alfalfa or fresh green grass is being fed, all additional nutrients can be met by increasing the amount of hay fed or pasture access.

Additional phosphorus is needed and may be supplemented in a mineral mix. However, the amount of hay or pasture forage fed cannot exceed the llama's maximal feed intake capacity (2% of body weight) and additional calories may be needed for some llamas, provided in a denser feed such as pellets or grains. If grass hay is fed, mineral mix should be provided (50/50 dicalcium phosphate and salt) as well as 1/2 to 2 pounds of grain per day. The grain quantity depends on the amount and quality of hay or pasture and the general condition of the llama. Be careful not to overfeed the pregnant female as it may increase the chance of a difficult delivery.

3. **The Lactating llama:** Lactation greatly increases the llama's requirements for nutrients. Generally, feed as above for late gestation. Mature alpacas rarely need grain supplements, but heavily lactating females (and rapidly growing youngsters)

may need the extra calories provided by grain.[12] After three months of lactation, gradually decrease the mother's feed to a maintenance level. At the time of weaning, stop feeding grain and decrease the amount of hay fed. This decreases milk production and prevents excessive distension of the mammary gland.

4. **The Nursing Baby:** If the cria, or baby, and its mother are doing well, it is best not to feed the baby grains or pellets until it is at least three months old. A normal llama cria should gain 1/2 to 1-1/2 pounds per day during this time while a normal alpaca cria should gain about half this amount. Most babies will start nibbling on solid food by three weeks of age. After the third month of lactation, the mother's milk production is decreasing while the baby's nutritional needs grow. The baby will compensate by eating with its mother. A free choice mineral supplement with phosphorus should be provided for the growing baby.

5. **The Bottle Fed Baby:** Colostrum (mother's first milk) is essential within the first 24 hours of life (preferably within eight hours). The baby needs 5 to 10% of its body weight in colostrum in its first day (which is probably all of its mother's colostrum). Thus, a 30-pound baby needs about one and a half to three pints of colostrum (one pint = one pound). This should be administered only if the baby cannot get it from its mother by itself. When administering this colostrum, divide it into 3 to 6 feedings during the first day of life but ideally within 12 hours. If llama colostrum is not available, use goat colostrum. When using goat colostrum, try to obtain it from a goat vaccinated against enterotoxemia and tetanus, and free from disease. Powdered cow colostrum replacers such as Colostrix® or Immustart® are now available through feed and animal health supply stores. These may provide an emergency backup when other options are not available. Colostrum can be frozen and

saved for emergencies. Do not thaw it in the microwave oven, as this breaks down the important protective antibodies in the colostrum. The colostrum should be between room temperature and 100 degrees F when it is fed.

After the first day, the baby llama can be fed 10% of its body weight of goat milk or Land-o-Lakes® milk replacer (1 part replacer: 6 parts water[13]) divided into about 8 to 12 feedings. The milk can be given with a regular baby bottle, but many options for bottles and nipples are available. The hole in the nipple may need to be enlarged in regular baby bottles to about 3 mm, but avoid large holes as they allow too much milk into the baby's mouth and may lead to aspiration pneumonia. Make sure the bottle, nipple, and your hands are kept clean, so as not to cause an infection in your baby llama. You may need to unscrew the cap as the baby drinks to keep constant pressure in the bottle.

The baby can be switched to a creep feed ration of 16% protein grain mix at 2 to 2-1/2 months of age. A creep feed is a balanced ration offered free choice to the babies only. Since the babies are accompanied by their mothers, the feed must be placed in a manger inaccessible to the adults. A room or stall with a small entry door works well. The babies can come and go as they please, but the adults cannot fit through the door.

The Berserk Male Syndrome

The berserk or aberrant male syndrome must be mentioned in regard to the bottle-raised baby. This is an abnormal behavior which results when a baby male llama bonds to humans rather than llamas. Bottle feeding and raising a baby male llama away from its mother is the most common cause. It can also be caused by excessive human contact with a baby even when it is still with its mother. When the male llama reaches puber-

ty, it develops aggressive behaviors which are normally directed towards other llamas. However, in berserk males, it is directed toward humans. Once berserk, the llama is dangerous and will not have a normal respect of humans. The change is permanent, and once expressed cannot be remedied by training or castration.

To avoid berserk male syndrome, allow bonding to occur between the mother and her baby. Interfere ONLY when necessary. If the baby must be bottle fed, allow it to stay with its mother and/or other llamas, when possible. Keep human contact to a bare minimum.

Most llama breeders agree that bottle-raised males should be neutered. Neutering is not a guarantee to prevent berserk males, but it may help. The llama should be neutered at an early age and before puberty. The best age for neutering these llamas differs among llama breeders but many neuter them between 2 and 6 months of age. In general, do not interrupt nature by excessively handling these babies. Baby llamas are meant to bond to other llamas, not to cuddling humans.[14]

Determining Ideal Llama Condition for the Average Llama

Due to the llama's thick coat, it is difficult to assess the llama's weight and fitness. A scale is invaluable for monitoring the weight of your llama. Weight loss may be indicative of illness. Progress on dieting llamas can be monitored too. Also, drug dosages can be accurately calculated based on the animal's true weight. The following are rules of thumb in determining llama condition.

1. It is normal to be able to feel the ribs easily through the fiber. If you cannot feel the ribs, your llama is too fat.
2. The inner thighs are bare of fiber and easy to see. If they jiggle while your llama walks, it is too fat.
3. The breastbone should be easily felt and

seen between the front legs. If no bone can be felt and the chest is the consistency of Jell-O®, your llama is too fat.

4. Feel across the back of your llama, just behind the withers. If it feels well-padded and not bony, your llama is too fat. Conversely, if all you can feel are bones across the back, your animal may be underweight.

5. Finally, do not determine condition according to the pelvic bones. Even fat llamas have very bony, prominent pelvic bones.

Hazards of Obesity

1. Fat llamas, just like fat humans, may have heart and circulation problems, and a decreased life span.

2. Fat llamas are more prone to hyperthermia (overheating).

3. Fat llamas may have decreased fertility and sex drive.

4. Fat female llamas have a higher incidence of dystocia. Fat accumulates along the pelvic canal and decreases the area necessary for reproductive examinations or birthing.

5. Fat llamas are prone to poor milk production. Most commonly, poor producing females tend to be fat with excessive fat stores in the mammary glands.

6. Llamas and alpacas in North America are often overfed and easily may become obese.

Llama Diets for Weight Reduction

1. Discontinue free choice feeding (including grazing)

2. Feed no grain.

3. Feed grass or oat hay with the grain seeds shaken loose.

4. Feed about 1% of your llama's body weight per day.

5. Provide a mineral and salt mix, and have water available free choice.

6. Increase exercise, but don't over-exert fat llamas, as they are more pone to hyperthermia.

7. Rebreed fat mothers as soon as possible after birthing.

8. Wean fat mothers' babies later than normal.

9. Weigh the llama often to monitor the program.

10. Fat females should not be placed on a diet during late pregnancy or during lactation. However, avoid overfeeding so as not to exacerbate the problem.

NOTES

CHAPTER 5:
HERD HEALTH

Introduction

A herd health program is basically a disease control program. Each farm needs a different program depending on the location, available feeds, animal density, contact with other species, and housing facilities. A precise program should be worked out with your veterinarian to fit your specific needs. The following are general areas to consider when designing your herd health program.

Records

The cornerstone of any herd health program is current and accurate records. Records are essential in determining routine vaccination and deworming schedules. They are extremely important in the diagnosis of infertility cases, herd disease outbreaks, and for determining heritability of certain traits. Each INDIVIDUAL animal must have its own record. A folder may be used with the following information on the front: animal's name, color, markings, birth date, sex, and registration number. If the animal was not born on your farm, the farm of origin and date of purchase should be noted. Purchase price may be included if desired. Inside the folder, a separate page or section should be kept on the following categories:

1. **Pedigree:** As much of your animal's pedigree as possible should be included. Color, size, and any abnormalities of relatives should be documented if known.

2. **Fiber:** This includes the animal's fiber quality and length. List any abnormalities such as bald spots.

3. **Housing, feed and show schedule.**

4. **Weight and height:** This is the weight and height of the animal when born, at weaning, and as an adult. The height is measured from the ground to the withers. Horse measuring sticks work well and are available at tack and feed stores. The purchase of a good scale is highly recommended. Although new scales are costly, used scales may be purchased at a reasonable price through want ads and livestock auctions.

5. **Vaccination:** These records include the dates and the products used for each vaccination. Note any adverse reactions to the vaccinations such as fever or swelling.

6. **Parasites:** Names of any diagnosed parasites and fecal examination results should be recorded. Dates and products of all dewormings and any external parasite therapy should be included. Again, any adverse reaction to medications should be noted.

7. **Teeth:** This category includes dates of fighting teeth removal and the method of removal. Any other dental problems such as tooth loss or tooth infection should be noted.

8. **Feet/legs:** This includes dates nails were trimmed, injuries, lameness, and angular limb deformities (crooked legs).

9. **Illness/injury:** The nature of any illness, as well as the therapy must be noted.

10. **Reproduction:** Any difficulties in breeding and birthing or any genetic concerns should be noted. Any reproductive problems in your animal's parents, siblings, or offspring should be listed. For the female llama, include the following: dates bred and to whom; length of gestation; whether the delivery was normal; and baby birth weights of her babies. If possible, include the offspring's adult weight and height, length of time until the placenta passed, mammary gland abnormalities, and previous infertility or medical problems. For your male, include dates he bred a female and whether a live baby resulted, sex of offspring, any semen evaluations, and any infertility problems. Record the date of castration if applicable.

11. **Castration:** Record the date of castration, if applicable. Castration, done as a surgical procedure by a veterinarian, is indicated for animals that are not breeding quality and those that are overly aggressive. While castration can be done at any age, waiting until the male is 1-1/2 to 2 years old allows maximum bone maturity. Castration done earlier than this, may lead to abnormal bone growth resulting in a tall llama with an increased tendency towards skeletal problems.[17] However, castration should be done earlier with bottle-raised crias as described in Chapter 4.

While this record system may seem very detailed, the information is invaluable to your veterinarian in diagnosing and treating your llamas. If you sell your animal, these records provide new owners with a specific medical history. The records will also help you to determine the best match for breeding your animal and will help you and the llama/alpaca industry discover and reduce the perpetuation of harmful genetic defects.

Nutrition

Nutrition is also in integral part of herd health. In general it is helpful to keep records of types of feed and any supplements used. See Chapter 4, as an entire chapter is devoted to nutrition.

Parasites

Several different species of parasites are known to affect llamas and alpacas. Parasites contribute to fiber damage, bald areas, dermatitis (skin inflammation), colic, weight loss (which is sometimes severe), diarrhea, constipation, anemia and general unthriftiness. Most intestinal parasites are passed through the feces and a llama ingests (eats) the parasite in its egg or larval form. Thus, if one llama has parasites, it is likely that other llamas in the same pen are also infected. Therefore, cleanliness is essential in parasite control. Your veterinarian should run periodic fecal examinations to check your llamas for worms and make frequent skin examinations to check for external parasites.

Life Cycles of Some Common Llama Parasites

A. Internal Parasites

1. ***Trichostrongylus*** (Stomach worm)

   The eggs of this parasite are passed in the feces of the llama. These eggs hatch and develop into larvae in the soil and in the feces. Then, a llama may eat these larvae on its feed or from the water, or directly from the soil. The larvae grow to adulthood in the stomach and small intestine of the infected llama. Eggs are then laid by the adult parasite in the digestive tract and passed into the llama's feces. This continues the parasite life cycle.[1]

2. ***Nematodirus*** (Thread-necked strongyle)

   The eggs of this parasite are passed in the llama's feces. They are very large and very distinctive eggs upon microscopic examination. These are sturdy eggs and may even survive the winter in the feces or soil. The infective larvae develop inside the eggs.

The eggs hatch and the larvae may survive for several months in the soil and vegetation. A llama then eats the larvae on ingested forage. The larvae mature into adults in the llama's small intestine. The adults then lay eggs and start the cycle all over again.[2]

3. *Trichuris* (Whipworm)

This parasite's eggs are passed in the feces of the infected llama. The eggs have a distinctive shape as if there were clear plugs at each end when examined microscopically. The infective larvae develop inside of the eggs. These infective eggs may survive for years. The eggs are eaten by a llama, and they grow to be adults inside the llama's cecum and large intestine. The adults then lay eggs to continue the life cycle.[3]

4. *Strongyloides* (Threadworm)

This parasite's eggs are passed in the llama's feces. The eggs are distinctive in that they contain a larva when laid. The larva then hatches on the vegetation or soil. The larvae can penetrate the skin of the host llama or enter via the mouth. After some migration in the llama's body which frequently includes the lungs, the larvae reach the small intestine and mature into egg-laying adults. Once the llama is infected, it takes about one week for the larvae to mature and begin laying eggs.[4]

5. *Moniezia* (Tapeworm)

The adult tapeworm is an off-white flat and segmented worm. It attaches to the wall of the llama's small intestine. Segments of the tapeworm containing eggs are passed into the feces and may be visible with the naked eye. The eggs are then eaten by an intermediate host, the oribatid mite. The mite, which lives on the grass or on feces, contains the tapeworm larvae and is then eaten by the llama. The larvae are freed during digestion and attach to the llama's intestinal wall where they mature and continue the life cycle, but generally cause minimal damage.[5]

6. *Eimeria* (Coccidia)

Coccidia are tiny protozoan (one-celled) organisms which can cause disease in many animal species. The outbreaks of disease are more common in the fall and winter months. The disease is also more common in young animals or new arrivals on your farm. The source of infection is the feces of an animal which has the disease or is a carrier. A carrier spreads the organism, but shows no signs of illness itself, thus it can be a continuous source of infection.

The feces may then contaminate the feed and water supplies. This most commonly occurs when the animals are crowded into small pens or where unsanitary conditions exist. A susceptible animal then ingests the organism. However, a very large number of coccidia organisms must be ingested in a very short period of time in order for the animal to get sick. If the animal only ingests a few organisms, there are no signs of disease. Repeated infection with small numbers of organisms produces immunity without clinical disease.

If the disease is present with clinical signs, the primary sign is diarrhea. If the disease is mild, the feces are watery and the animal is only slightly depressed for three to four days. Severe cases have diarrhea with blood in it, the animal becomes depressed, anemic, strains repeatedly, loses weight, may have convulsions, and could die.

A recovered animal has immunity to that particular species of coccidia. A routine fecal exam for coccidia should be done whenever several animals are kept crowded in the same pen. If you have one develop diarrhea, the sick animal should be isolated from the others, and the veterinarian called for diagnosis and therapy.

If coccidiosis is an ongoing clinical problem in your herd, preventative therapy can be used. The most important part of prevention is sanitation. Over-crowding should be avoided and stress should be minimized. Feed and water containers should be high enough to prevent fecal contamination. Pens should be kept as clean as possible.

Preventative medicines are also commercially available (Corid®, Deccox®, Bovatec®), and should be used when pen crowding is at its worst. These medicines are added to the drinking water. To be effective, all other water sources should be eliminated, so that the medicated water is all that is available to the group of llamas.

7. *Parelaphostrongylus* (Meningeal worm)

Parelaphostrongylus is a natural parasite in white-tailed deer. This means that no disease occurs in the deer. The worm is passed from an infected deer through its feces, then develops in a snail which subsequently may be ingested by another deer. The llama or alpaca may become an accidental or abnormal host when it inadvertently eats a snail or slug containing the infective worm. *Parelaphostrongylus* migrates along the central nervous system thus causing neurologic abnormalities in the llama. Signs include uncoordination, paralysis, knuckling or dragging of limbs, blindness, and the disease may worsen to the point of death. If your llama is showing neurologic ahnormalities, and you are in an area populated by white-tailed deer, consider the meningeal worm as a possible cause of the problem and contact your veterinarian for advice on treatment and control.[6]

8. *Fasciola hepatica* (Liver fluke)

The adult liver fluke is a flat leaf-shaped organism found mainly in wet and swampy areas where snails live, such as the southern and southeastern regions of the United States. The adult fluke lives in the bile ducts of its host's liver. The adult lays the eggs in the bile ducts and the eggs are passed out in the feces of the infected llama. The eggs develop and hatch in water and live as a free-swimming organism. The intermediate host, the snail, then becomes infected with the free-swimming organism. The larval stages continue in the snail and break out of the snail to emerge as another free-swimming organism. They attach to the vegetation and become encysted. Then, they are eaten by a llama and penetrate its intestinal wall to migrate through the liver for about six weeks. Next, they enter the bile ducts to mature and lay eggs, which takes another couple of weeks.[7]

In the llama, the flukes can cause liver damage, bleeding resulting in anemia, protein loss, loss of appetite, and digestive upsets.[8]

9. *Eperythrozoon or Mycoplasma*

Eperythrozoonosis (Epe) is caused by a blood parasite called *Eperythrozoon*, more recently named *Mycoplasma*, and has been diagnosed in llamas. This parasite can be transmitted from one animal to another by biting insects, blood transfusions and reused hypodermic needles. *(This reinforces the absolute need for use of new or sterilized needles for each llama when administering injections.)* Signs of infection with this parasite range from no signs, to fever, anemia, weight loss, lethargy and weakness.[9] Since actual illness is rarely seen in other species, it is possible that *Eperythrozoon* caused illness in llamas may be secondary to other disorders such as immune system problems.[10] Eperythrozoonosis should be considered in any poor-doing llama. Blood tests are used for diagnosis and treatment with antibiotics can begin upon a positive diagnosis.

B. External Parasites

1. *Sarcoptes* (Scabies)

Sarcoptic mange is caused by *Sarcoptes scabeii*. This is a mite whose entire life cycle is spent on the animal. The mite burrows into the outer layer of skin. It prefers the tender skin areas with thin hair coats such as the face, belly, chest, and legs. The area initially develops hairless spots, dandruff, scabs, and crusts. Itching may or may not be severe. As the disease continues, the skin becomes thickened, crusty, and leathery. The disease can be spread among llamas and other animals. If your llama has skin problems as above, call your veterinarian to do a skin scraping to look for the offending mite. Currently, ivermectin injections are used in treatment of mange.

2. *Psoroptes*

Psoroptic mange is caused by a tiny, white-colored host-specific mite called *Psoroptes communis*. They occur almost universally and are found year-round but become more common in the cold months. They live entirely on the animal's skin feeding on serum. They prefer heavily fibered areas and can cause intense itching. Infestations are diagnosed by a skin scraping done by your veterinarian who can suggest treatment options. Due to an eradication program in the United States, your veterinarian will need to report the presence of this mite to the state authorities.

3. *Chorioptes*

Chorioptic mange occurs universally and infestations are most common in the winter. These mites live in colonies on the skin surface and prefer the legs and underside of the llama. Signs resemble Sarcoptic mange but are usually not as severe. Your veterinarian can diagnose this mite by skin scraping and treat the infestation appropriately.

4. *Lice*

Lice are small (2 to 6 mm) wingless insects. Two types of lice can infest llamas. One is *Mallophaga* which is the chewing louse. Chewing lice nibble on hair and debris on the llama's skin surface. They are a nuisance to the animal because they cause skin irritation. The other type of louse is *Anaplura,* the sucking louse. These can be distinguished microscopically from chewing lice by their narrow heads. Sucking lice feed entirely on blood and can therefore cause anemia (blood loss), spread disease, or cause an allergic reaction by their presence.

Lice infestation can be passed between llamas by direct contact or by shared tack, grooming equipment, feeding and housing areas.[11] Signs of lice infestation include restlessness, scratching, rubbing the affected areas, increased dandruff, ragged fiber, and shedding of fiber in large patches. Infestations tend to be worse in the winter months. Diagnosis can be made by examining the fiber and skin for the presence of the lice or their eggs (nits). The nits are light in color and about 1 mm long. They attach firmly to the hair. A blower can help part the fiber for ease of examination. The treatment includes an ivermectin injection for the sucking louse and topical carbaryl dust or fenthion pour on (Tiguvon®) for the chewing louse. (Use the swine dilution for fenthion.)[12]

5. *Ticks*

Many different species of ticks can infest llamas. Tick type is generally dependent upon geographical area. Adult ticks and immature stages feed on blood. Ticks could cause a large enough blood loss to cause anemia. While feeding, they release an anti-clotting substance to facilitate feeding. This substance can also cause an allergic type of irritation to the skin. Ticks also carry diseases which can be passed on to the llama via the bite (e.g., Bluetongue and Anaplasmosis could be transmitted this way).

Signs of tick infestation may include weakness, poor appetite and infected skin areas. Shaking of the head is present from ear ticks. Also, a localized swelling, which can be quite large, may be present from the tick bite directly or from a secondary infection. Signs will depend on the type and location of the tick and the number of ticks infesting your llama.

Some ticks can release a toxin (poison) in their saliva that can cause paralysis of the animals they are biting. The disease starts with incoordination of the animal's back legs, leading to paralysis of the legs. As the disease progresses, the diaphragm or breathing muscle gradually becomes paralyzed. This causes difficult breathing. The breathing rate decreases, and when breathing becomes impossible, the result is death. Since the disease is potentially fatal, it is important to start looking for the offending ticks early in the course of this disease. Early removal of ticks causes a very rapid recovery. One to several ticks may be involved. The llama may need to be clipped in order to locate the ticks, however, the ticks tend to seek the tender skin areas (e.g., the sheath, mammary glands, and where the legs attach to the body.) Treatment with an external insecticide and possibly ivermectin may be needed to rid the llama of any ticks that are missed upon examination.

6. Nasal bots

Oestrus ovis, the sheep nasal bot, and *Cephenemeya* spp., the deer nasal bot, have been reported in llamas.[13] The adult insect is a fly resembling a bee. The female deposits her larvae (immature stage) around the llama's nose. The larvae then crawl up the nose and live in the nasal passageways and sinuses for two weeks to nine months. Development continues to the pupa stage, which drops back onto the ground and develops into an adult fly. The larvae in the nasal cavities and sinuses cause irritation which causes a runny nose. In the beginning the nasal discharge is clear, and later becomes yellow-white and could even be bloody. This also can cause sneezing, congested breathing, and rubbing of the nose. The more larvae present, the more signs of disease. Treatment includes ivermectin injection SQ, perhaps at twice normal dose.

7. Flies

The presence of flies goes hand in hand with raising animals. Flies are primarily an annoyance to animals and people but may cause other problems such as eye irritation from feeding on tears. Some flies such as Stable flies, Biting midges, and Buffalo gnats bite the llama and suck its blood, and the bite is very painful to the llama. The fly bites are irritating to the skin, and may cause itching and/or an allergic reaction. The amount of blood lost to these flies is rarely enough to cause anemia, but over a period of time may cause a loss of condition and weight. Flies can also act as a vector for carrying diseases from one animal to another.

In warmer climates such as the southern part of the United States, flies may be active for the entire year. In areas with a cold season, flies are present seasonally in the late spring through early fall. Most flies are active during the daylight hours. However, some of the tinier species of flies such as the Biting midges are most active at dusk and tend to be seen in swarms.

Most flies need animal manure on which to lay their eggs. For example, the Stable fly must lay her eggs in decaying organic matter such as manure. Some flies such as the Biting midges and small Buffalo gnats lay their eggs in moist areas such as rocks and plants that are splashed with water. Since these flies are very annoying to animals and people alike, the following are some suggestions to keep flies at a minimum at your llama farm.

a. Manure removal is the most important aspect of fly control, since so many of the flies need the manure for their eggs.

b. Avoid grazing llamas in pastures with swampy areas and ponds if possible. Gnats, midges and mosquitoes are more common in these moist areas.

c. Numerous insecticides for livestock are available for external parasites. Topical insecticides are available as powders, sprays, roll-ons, pour-ons, or liquids that are rubbed onto the animal. **Insecticides are poisonous and should always be used with caution.** Toxicities have been reported in llamas after insecticide treatment.[14] It is advisable to consult with your veterinarian before using any insecticide to determine its safety factor, whether it causes any adverse reaction or odors to the fiber, and whether it can be used in pregnant, lactating, or young llamas.

Some llamas, especially dark faced llamas, are particularly bothered by flies around the face and eyes. These llamas would benefit from fly repellent rubbed on the face and around the eyes. The pyrethrin based roll-on for horses available at feed stores works well, is easy to apply, and is very helpful to take along on pack trips. Likewise the short-haired areas on the legs could benefit from some fly repellent if the llama is greatly bothered.

Alternatively, collars or tags which release Ectrin® insecticide can be used to aid in pest control around the face. The tags can be placed on the neck with an improvised breakaway collar that will break if caught on an object.

d. Fly traps can be used to help control fly problems. Bait such as meat is put in special jars available through most animal feed and supply stores. The decaying meat attracts the adult flies and the lid is constructed so they cannot leave. Thousands of flies can be trapped per jar by this malodorous method.

e. Electronic insect killers use light to attract and electrocute flying insects. Although effective for insects in general, flies are often more attracted if bait such as decaying organic matter is used in association with the "zapper".

f. Biological controls such as parasitic insects will also help control fly populations. Immature forms (pupae) of these tiny stingless insects are distributed onto the llama manure where many flies develop from eggs to adults. These harmless parasitic insects feed on and kill the immature flies. They subsequently develop into adult insects and lay their eggs on manure piles to continue the cycle of parasitization.

Table 1
Internal Parasites of Llamas

Stomach
> *Haemonchus* (Stomach worm, Barber pole worm)
> *Ostertagia* (Stomach worm)
> *Camelostrongylus* (Stomach worm)
> *Spiculopteragia*
> *Physocephalus*
> *Graphinema*
> *Gongylonema* (Gullet worm)

Small Intestine
> *Ostertagia*
> *Trichostrongylus*
> *Nematodirus*
> *Cooperia*
> *Capillaria*
> *Moniezia* (Tapeworm)
> *Eimeria* (Coccida)
> *Thysaniezia* (Tapeworm)
> *Lamanema*
> *Bunostomum* (Hookworm)
> *Giardia*

Large Intestine
> *Trichuris*
> *Oesophagostomum*
> *Chabertia*
> *Skrjabinema* (Pin worm)

Lungs
> *Dictyocaulus* (Lungworm)

Liver
> *Fasciola* (Liver fluke)
> *Dicrocoelium*

Nervous System
> *Parelaphostrongylus* (Meningeal worm)

Eye
> *Thelazia* (Eyeworm)

Blood
> *Eperythrozoon* = *Mycoplasma*

Murray Fowler, D.V.M., "Proceedings from Llama Medicine Workshop for Veterinarians," 1987, pp. 44-49., C.S.U., Fort Collins, Colo.

Carl E. Kirkpatrick, V.M.D., Ph.D., "Giardiasis in Large Animals." Compendium on Continuing Education for the Practicing Veterinarian, Vol. II, No. 1 (January 1989), pp. 80-84.

Table 2
External Parasites of Llamas

Mites
> *Sarcoptes scabeii*
> *Psoroptes communes*
> *Chorioptes*

Lice
> *Mallophaga* (chewing)
> *Anaplura* (sucking)

Ticks
> *Amblyomma* (Lone star tick, Gulf coast tick)
> *Rhipicephalus* (Brown dog tick)
> *Dermacentor* (Rocky Mountain wood tick, Spotted fever tick)
> *Otobius* (Spinose ear tick)
> *Ixodes* (Black legged hard tick)

Bots
> *Oestrus* (Sheep nose bot)
> *Cephenemeya* (Deer nose bot)

Flies
> *Tabanus* (Horsefly, Deer fly)
> *Musca domestic* (House fly)
> *Musca autumnalis* (Face fly)
> *Stomoxys calcitrans* (Stablefly)
> *Calliophoridae* (Blowflies)

Mosquitoes
> *Culex*
> *Anopheles*
> *Aedes*

Fleas
> *Vermipsylla*

Murray Fowler, D.V.M., "Proceedings from Llama Medicine Workshop for Veterinarians," 1987, pp. 44-49, C.S.U., Fort Collins, Colo.

General Parasite Control Guidelines

In general, the following will help control parasites:

1. Deworm ALL llamas on the premises whenever deworming takes place. The interval of deworming depends on the density of your llamas. It may range from

one to six times per year. Consult with your veterinarian for the deworming frequency on your particular farm, and for the product to be used.

2. Rotate pastures, if possible, allowing a minimum of one month pasture rest.

3. Crowding and overgrazing should be avoided.

4. Feed and water supplies should be away from manure piles, and constructed so that they will not become contaminated with feces. Clean tanks and feeders regularly, and keep feed OFF the ground.

5. Manure should be removed regularly.

6. Fence off wet and swampy areas in your pasture.

7. Llamas share some of the same parasites with sheep, cattle, and other ruminants. Therefore, grazing them together should be avoided.[15]

Vaccination Programs

An important part of disease control is a regular vaccination program. Although developed for usage in other species, such as sheep and cattle, many vaccines have been used in llamas and alpacas. Which you use depends on your specific geographic area and farm environment. Consult with your local veterinarian as to which diseases are a problem in your area. Some general guidelines for vaccinating your llamas and alpacas are as follows:

1. Adequate restraint for safety and ease of administration is a must.

2. If administering your own vaccine, remember to keep the vaccine refrigerated, and to use a new needle and syringe with each animal.

3. Adequate antibody or protection level usually requires a booster shot three to four weeks later in a previously unvaccinated animal. Adequate response to the vaccine takes two to four weeks following the vaccination.

4. Vaccination of healthy animals results in maximal antibody response. Avoid vaccinating animals that are ill or under stress if possible.

5. Adverse vaccination reactions such as swelling, excessive muscle soreness, respiratory difficulty, or shock, should receive veterinary attention and be recorded in your files. Always consult with your veterinarian about the safety of readministering a vaccine that resulted in an adverse reaction.

Frequently Used Vaccines

Enterotoxemia

Enterotoxemia is caused by a bacterial microorganism named *Clostridium perfringens.* This is a SEVERE intestinal disease caused by a toxin or poison made by several types of this Clostridial organism. These bacteria are normally found in soil and in the intestine, but under certain conditions grow to high numbers and produce toxins. These conditions include rich diet, sudden diet changes, and overeating. The disease has a very rapid course often resulting in sudden death before any signs are seen. If seen before death, these llamas are very depressed, may have diarrhea, may be uncoordinated, weak, in pain, have convulsions, or be in a coma. The prognosis is very poor. The enterotoxemia type C and D vaccine is a MUST in your herd health program. Current recommendations are to vaccinate all llamas yearly. To minimize the risk of a stress-induced abortion, give pregnant llamas their boosters prior to their last two months of pregnancy. Vaccinate babies at one and two months of age, followed by yearly boosters.

Tetanus

Tetanus is also caused by a Clostridial organism called *Clostridium tetani.* The organism resides in the soil and in animals' intestinal tracts. The bacteria do not like oxygen, and thus grow well in deep non-aerated puncture wounds. The bacteria then release a toxin that affects the nervous system and causes muscular spasms. Spasms of the chewing muscles can cause the characteristic "lockjaw." Tetanus vaccine is a MUST in your herd health program. Products are available that have Enterotoxemia Types C and D plus tetanus in the same vaccine. Recommendations are to

49

vaccinate all llamas yearly. Boosters to pregnant females may be done concurrently with the enterotoxemia boosters. Vaccinate babies at one and two months of age followed by yearly boosters. Give a booster following deep wounds.

Malignant Edema

This is also a Clostridial disease caused by *Clostridium septicum*. The organism resides in the soil and in animal's intestinal contents. Disease may result from dirty wounds, snake bites, or, rarely, following contaminated injections. The llama becomes very ill with fever and severe swelling around the wound which often extends far beyond the original site. The disease is often fatal. Consult with your veterinarian as to whether your area has a high risk factor for malignant edema as shown by a history of malignant edema, snakes in the area, or a high incidence of wounds in the herd. In high risk areas, this vaccine is a must, and highly recommended elsewhere. Vaccinate all llamas yearly and give a booster following a severe injury.

Blackleg

Blackleg is caused by *Clostridium chauvoei*. The organism can be found in the soil and intestinal tract of animals. The organism travels from the intestine to the bloodstream and then to the muscles. Infection flourishes in the muscles secondary to bruising or may even occur spontaneously. The animal becomes feverish with swelling over the muscle. The disease rapidly progresses and is often fatal. Sometimes, the only sign is sudden death of the animal. Yearly vaccination is recommended in high risk areas (as above). The vaccine is available in combination with the other clostridial vaccines, including malignant edema.

Rabies

Rabies is caused by a virus. All warm blooded animals can be affected by the rabies virus. It is typically transmitted from a bite of a rabid animal. The saliva contains the virus, which then enters the animal through the bite wound. The virus then travels along nerves to the spinal cord and brain, where the virus multiples. From there, it travels along nerves to the animal's saliva, thus making the saliva contagious. The clinical signs of rabies vary from depression, paralysis, to extreme anxiety, agitation, and viciousness. Excessive salivation or frothing is also typical. The "carrier pool" of rabies depends on your specific geographical area. Skunks, bats, dogs, cats, foxes, coyotes, bobcats, and raccoons are common carriers. Vaccination against rabies depends on the occurrence of rabies in your area. If you in an endemic (rabies prevalent) area, have your veterinarian vaccinate your llamas yearly for rabies with a killed vaccine.

Leptospirosis

Leptospirosis is caused by bacteria which can infect people and animals. The organism survives well in water and is most commonly found in moist areas (surface ponds, stagnant water). The bacteria enters the body by intake of contaminated water or food. It also can enter the body through the skin and mucous membranes (mouth, genitals). The organism then localizes in the kidneys. The animal can become a carrier of the disease and shed the organism in its urine for long periods of time. Signs of the disease vary from few and mild signs, to severe signs. The severe disease signs include bloody urine, jaundice (yellow color to skin and membranes), fever, depression, anorexia (no appetite), diarrhea, and abortion.

The need for vaccination depends on your area and the level of cleanliness on your farm, and the types of areas to which you and your llama travel. Your veterinarian can best advise on the need for this vaccination. Vaccination frequency ranges from one to two times per year, again depending on the problem potential in your area. Fencing off ponds and stagnant water will also decrease the chance of infection. If your animal is diagnosed with Leptospirosis, consult with your veterinarian for isolation and cleanliness procedures, as it can be passed on to people and other animals.[16]

West Nile Virus

See Chapter 15 for information on seasonal vaccination for West Nile virus.

General Disease Control and Prevention

Herd health also involves some common sense. Not all diseases can be prevented by vaccinations. If you have a sick animal, keep it isolated until your veterinarian determines if the illness is contagious. All new animals added to your herd should be isolated for at least four weeks before joining your herd. New animals may look healthy upon arrival, but may be incubating or carrying an infection. Also, when llamas are exposed to numerous other llamas at a show or event, it would be advisable to isolate your llama for four weeks after returning home. Your llama may be just about to come down with an infection.

Another important consideration is the number of animals you have in a given area. If the population is highly concentrated, the chance of spreading infection or parasites greatly increases. A less dense herd will tend to be healthier.

Finally, ventilation must be adequate in barns and trailers. If ventilation in the barn is poor, the number of disease-causing organisms greatly increases. This, plus the irritation from the accumulated dusts and fumes, makes the llama more susceptible to respiratory infections such as pneumonia. If trailers are poorly ventilated, disease can be quite severe because the llama may also be hyperthermic and stressed by the trip-which decreases its ability to combat disease.

With the above herd health management procedures, disease can be kept to a minimum. A complete vaccination program, good nutrition, parasite control, accurate records, common sense, and keeping stress to a minimum will keep your llamas and alpacas as healthy as possible.

NOTES

CHAPTER 6:
TEETH

Introduction

Adult llamas normally have three pairs of front teeth or incisors, located on the lower jaw. The intact male has three pairs of fighting teeth, two upper pairs and one lower pair. In the female, the fighting teeth are usually rudimentary. Males gelded at a young age also tend to have small fighting teeth. Next, llamas have one to two pairs of premolars on the upper and lower jaw, and three pairs of molars on the upper and lower jaw.[1] (Figure 6.1) The premolars and molars are the grinding teeth towards the back of the mouth. They are normally sharp and should not be confused with "points" or abnormally sharp edges. However, since they are sharp, use extreme caution when putting your fingers in the llama's mouth.

Llamas, like other mammals, normally lose their baby or deciduous teeth which are then replaced by permanent teeth. It can be very alarming for llama owners to find their llama with a bloody mouth, and a tooth in the feed manger! In general, the permanent central incisors erupt at 2-1/2 years, the middle incisors at 3 to 3-1/2 years, and the last or corner incisors at 4 to 6 years. The fighting teeth erupt at 2 to 7 years,[2] with average being 2-1/2 years. The fighting teeth are very large and pointed in the adult male llama. They can be used as harmful weapons against other llamas or people. For safety reasons, the fighting teeth should be removed or filed off in the male.

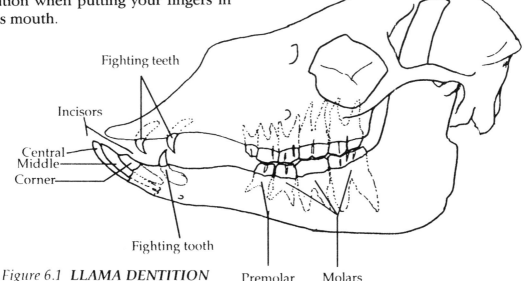

Fighting teeth

Incisors

Central
Middle
Corner

Fighting tooth

Figure 6.1 **LLAMA DENTITION**   Premolar   Molars

Fighting Teeth and Tooth Sawing

It is seldom necessary to remove fighting teeth on females or quiet geldings, and even a stud male may be able to keep his fighting teeth as long as he lives by himself and never has an accidental encounter with another male. However, if two or more males are kept together, all of them should have their fighting teeth cut to avoid torn ears, bites on testicles, and other injuries. Fighting teeth will continue to grow, so if they are cut soon after eruption, it is important to check from time to time to see how much they have grown. If they are too long, they should be cut again. Usually this only has to be done twice in a male's life.

Fighting teeth can either be removed surgically by a veterinarian, or quickly and easily be sawed off by you or your veterinarian, using obstetrical wire. Most llamas seem to find the procedure nonpainful and nonthreatening. Often they are bothered more by having someone's fingers in their mouth than they are by the actual process of tooth-sawing.

In order to cut fighting teeth, the animal must be restrained securely to keep his head steady. Putting the llama in the chute and cross-tying the head works well. Two people are needed to saw fighting teeth. Both should wear glasses or some sort of eye protection, since the cut tooth can fly off at a high speed in an unanticipated direction. Since both fighting teeth and molars are very sharp, it may be helpful for the person holding the mouth open (referred to as the "holder") to wear gloves. The holder should stand on the side opposite from where the tooth will be cut. That is, if one of the llama's left teeth is to be cut, the holder should stand to the right of the llama's head, reaching over and under the llama's nose to hold the lips back away from the teeth. This is important to prevent the llama's lips from getting hurt, to keep the llama from pushing the wire saw out of his mouth with his lips, and so that the person

doing the sawing can see what he or she is doing. The person doing the sawing may also want to wear gloves, unless the wire saw has sturdy handles. If handles aren't available, it is usually easiest to wrap the last few inches of the ends of the wire saw around the little fingers, while the center loop is held securely between the forefingers and thumbs. A 1-1/2 to 2 foot piece of wire is a comfortable length to work with. After the first time a section of wire is used to saw a tooth, it will kink and curl into small loops and circles. This doesn't mean that the saw is used up, it simply indicates that it has been used. If more than one llama needs his teeth sawed, change to a new wire between llamas to avoid the possible spread of eperythrozoonosis. (See Chapter 5.)

To saw the tooth, the holder keeps the llama's lips open, and the person doing the sawing should form a small loop in the wire and pass this around the tooth. The wire should be kept taut at all times. When the wire is around the tooth, one or two quick pulls parallel to the gum line will cut a groove into the tooth. After the groove is present, the rest of the sawing usually goes very quickly, and the tooth will come off within five seconds or so.

Once it is off, check the tooth stub carefully by feeling with a finger. If it is very sharp, it may be necessary to smooth it down, either with a small metal file or with a piece of fabric backed sandpaper wrapped around a finger. Sometimes there will be a small amount of bleeding when a tooth has been cut, often due to damage of the gum, but it will usually stop quickly.

This method can be used for both upper and lower fighting teeth. The one modification needed for lower fighting teeth is to be careful to avoid the thin fold of membrane on the tongue side of the lower tooth. If damaged during tooth sawing, this will hurt and bleed.

Overly long incisors due to jaw malalignment can be cut for cosmetic reasons and to aid in prehension of food using this

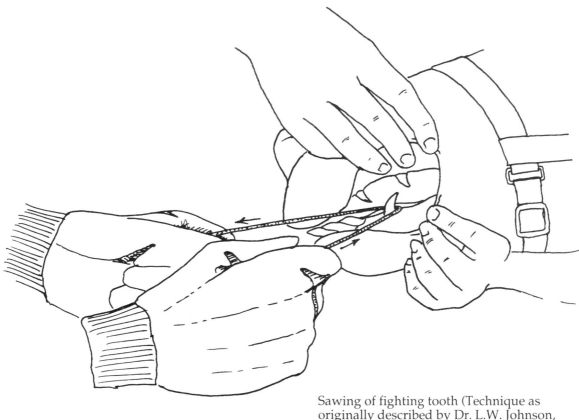

Figure 6.2 *FIGHTING TEETH REMOVAL*

Sawing of fighting tooth (Technique as originally described by Dr. L.W. Johnson, C.S.U., Fort Collins, CO)

same method. However, the teeth are thicker and heavier than the fighting teeth and the sawing will take longer. Also, the friction of sawing produces heat, and since there is a thick section to saw through, the heat produced will be painful. In this case, it is less painful for the llama if a third person can cool the tooth during sawing with a spray of cool water. Before sawing any incisors, consult with your veterinarian because sawing the ends from these teeth may expose the pulp cavity, which increases the likelihood of a tooth infection.

Retained Deciduous Teeth

As far as dental disease, llamas do not seem to have as many problems as we humans do. Occasionally, a deciduous or baby tooth will not have fallen out by the time the permanent tooth has erupted. The temporary teeth should be loosened gradually. This can

be done by rocking them back and forth gently over a period of several days. Then the tooth can be pulled out easily, if it doesn't fall out on its own. If this doesn't work, your veterinarian should pull the tooth.

Broken and Chipped Teeth

Sometimes llamas who are playing or fighting vigorously will knock out or break a tooth during sparring, which can cause a very bloody mouth. This occurs most commonly with deciduous teeth because their roots are not as extensive and secure as in permanent teeth. Young males may look quite piratical with several chipped or broken baby teeth! Luckily, the broken or missing deciduous teeth will be replaced with permanent teeth as the animal matures. Broken deciduous or permanent teeth should be examined to determine the extent of the problem. Loosened teeth may need to be removed, and the possibility of infection will need to be assessed by your veterinarian.

Infections

Other dental problems may be marked by the development of a firm, bony lump on the cheek or the jaw. A lump may just be a wad of food but should be examined as it is often indicative of a tooth infection. Any tooth and its root may become infected, but whenever there is damage to a tooth, whether it is broken or sawed off, the risk is a little higher. The bony lump forms in the area of the infected tooth or tooth stump. Infection can be confirmed by X-ray, and if present, it will need medical or possibly surgical attention. Antibiotics are used to combat the infection, but dental surgery to remove the bad tooth is usually indicated.

Dental Problems in the Older Animal

As the llama ages, old or infected teeth often loosen and fall out. When a llama is missing teeth or has irregular wear of the teeth, it will cause problems with the first step of digestion, the mechanical breakdown of food through chewing. Such a llama will begin to lose weight even if it is still eating enough food. Sometimes it will eat less food because the mouth may be sore. When faulty chewing is observed, an oral examination should be done by your veterinarian. It may be necessary to float or file irregular edges on the teeth. If the llama still has trouble chewing or is missing several teeth, it should be fed softened food which does not require extensive chewing. Alfalfa pellets, perhaps mixed with a little grain and softened with water into a gruel make an ideal supplement for those llamas who no longer chew well enough to get the nutrients they need from hay or pasture.

Figure 6.3 **ABSCESS OF LOWER FIGHTING TOOTH.**

56

CHAPTER 7:
EYES

Introduction

Normal llama and alpaca eyes are large and beautiful with long eyelashes. Most of these animals have brown eyes. Some dark-eyed animals may have a slight hint of blue when the eyes are examined closely in bright light. Occasionally, a llama or alpaca has one or both blue eyes which are presumably inherited as a recessive (subordinate) trait. Very little pigment to the eye (iris) or no pigment at all does not mean that the animal is blind. However, these non-pigmented eyes MAY be more sensitive to light and some of the damaging ultraviolet rays from the sun. Many animal breeders feel that very light-colored eyes are not aesthetically pleasing and should not be a trait selected for in a breeding program.

Normally, both eyes are held wide open and full of expression. Tearing should not be seen normally, except when windy conditions exist. When examining the eye closely in the bright sunlight, an interesting structure called the corpus nigrum may be observed. The corpora nigra are dark brown irregular nodules that extend from the top and bottom edges of the iris (the colored part of the eye). These dark nodules dangle over the pupillary opening. These are normal structures and would appear to aid the pupil in regulating the amount of light entering the eye.

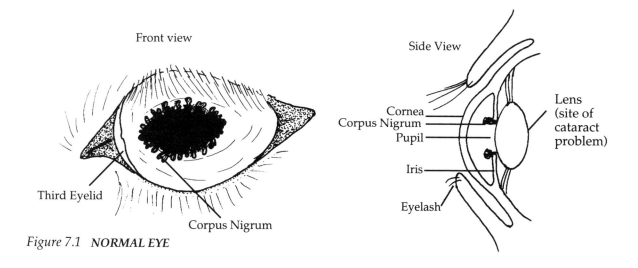

Front view

Third Eyelid

Corpus Nigrum

Side View

Cornea
Corpus Nigrum
Pupil

Iris

Eyelash

Lens
(site of
cataract
problem)

Figure 7.1 **NORMAL EYE**

General Eye Injuries

Unfortunately, eye injuries and infections do occur in llamas. They may be mild problems, or may result in scars, or, in severe cases, may result in blindness. The general signs of eye problems include a discoloration of the eyeball, blood or evidence of injury, excessive tearing, squinting, or holding the eye closed. Injuries to the eye or eyelid can occur due to encounters with sharp projections from fences, gates, sheds or tree branches. You may suddenly find that your llama has blood around its eye and face. These injuries often look worse than they are, as facial injuries do bleed a lot. If the injury is still bleeding, stop the bleeding by applying pressure on the area, preferably with a sterile dressing. However, by the time you find these injuries, they usually have already stopped bleeding. Next, you should examine the extent of the injury.

Figure 7.2 NORMAL AND ABNORMAL EYE.

Put your llama in the chute and then cross-tie it to stabilize its head. Flush the eyeball with sterile saline and gently wipe away the blood from around the eye. A syringe WITHOUT A NEEDLE works well for the flushing. (Figure 7.3) Now you can see if the injury is actually on the eyeball itself, the eyelids, or the conjunctiva. The conjunctiva is the pink membrane lining the eyelids and in the corners of they eyes. Sometimes a small flashlight can be used to shine light across the surface of the eyeball to highlight foreign bodies or damaged areas.

Eyelid Injuries

Large, deep eyelid injuries may go through the entire skin thickness and perhaps all the way through the eyelid. Such injuries will need to be sutured by your veterinarian. Do not use any ointment, or give any care other than rinsing the eye with sterile saline, until after the cut is sutured, and then follow your veterinarian's directions. Small eyelid injuries should also be checked by the veterinarian, for if they are not sutured, they may heal with non-cosmetic results. (Figure 7.4)

An unsutured injury may also heal so that the lashes turn inward, toward the eyeball, where they may cause eyeball irritation and damage. However, if the eyelid injury is small, and the edges are in line, it will usually heal quite well. Consult with your veterinarian regarding specific therapy. Most minor injuries will heal with minimal therapy in a few days.

Conjunctival Injuries

If, after flushing an injured eye with sterile saline, you notice that there is a scrape or blood oozing from a tear in the pink membrane lining the eyelid or the corners of the eye, your llama has suffered a conjunctival injury. If the injury is large, sutures may be needed. However, most injuries are small, and these heal very well when treated appropriately according to your veterinarian's instructions. Often ophthalmic ointments are prescribed by your veterinarian to be applied several times daily for a specified period of time or until the injury cannot be seen and excessive tearing and squinting have stopped. Put eye ointment into the eye by lowering the eyelid, and squeeze ointment into the "pocket" of lower lid. Don't touch the eye with the medication tube as it may spread infection to other eyes.

Llamas can also develop conjunctivitis. This term is used for inflammation of the eye membranes. This may range from a "pinkeye"-like syndrome which is often present in both eyes, to tired, sore eyes, caused by irritation from wind, flies, pollution, or fumes. Conjunctivitis may be associated with the pres-

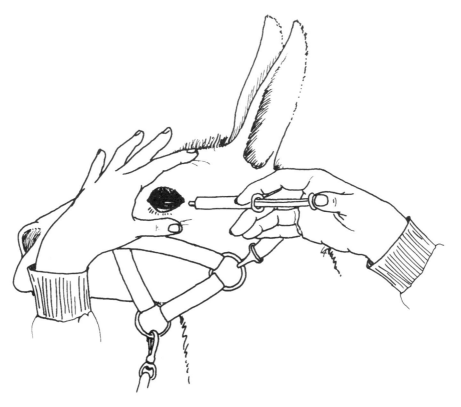

Figure 7.3 **FLUSHING EYE WITH STERILE SALINE.**

Figure 7.4 **NON-COSMETIC RESULT OF UNCORRECTED EYELID INJURY.**

ence of one or more nematode worms, *Thelazia californiensis.* These worms are found throughout the United States, not just in California as the name implies.

Conjunctival problems cause one or both of the llama's eyes to tear excessively. The llama may squint due to pain. Your veterinar-

dust, and providing a safe environment for your animals will go far to prevent conjunctivitis and other health problems.

Eyeball Injuries

The cornea is the transparent front surface of the eyeball itself. Corneal injuries and

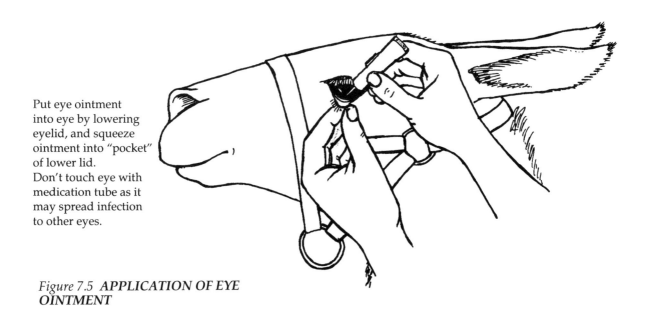

Put eye ointment into eye by lowering eyelid, and squeeze ointment into "pocket" of lower lid.
Don't touch eye with medication tube as it may spread infection to other eyes.

Figure 7.5 **APPLICATION OF EYE OINTMENT**

ian should examine the llama's eyes to determine the extent and cause of the conjunctivitis. Irritating foreign objects such as food material or wood splinters may be discovered and removed. The *Thelazia* parasite may be identified and removed or killed by your veterinarian. Additional therapy such as ophthalmic ointments or even surgery may be indicated. In some cases, removing the inciting cause may be the only treatment needed. For example, fly repellent rubbed around the eyes allows conjunctivitis caused by fly irritation to heal nicely on its own. With these mild cases, the excessive tearing and redness will usually disappear in three to four days. Obtain veterinary attention if there is no improvement in this period of time. Prevention is better than treatment whenever possible. Transporting your llamas in a closed trailer, out of the wind, taking measures to control flies and

ulcers are fairly common in llamas. The cornea is very sensitive, so whenever it is injured, it is quite painful. Ulcers are caused by irritation, infection, or injury to the eyeball. Because damage to the cornea is so painful, the llama will squint or hold its eye shut. Upon examination, you may be able to see a cloudy spot on the surface of the eye. This may be either an ulcer, a laceration (cut), a scar, or more rarely, an abnormal growth. Sometimes, the injury to the cornea is so small that it cannot be seen easily. Any painful or abnormal eye as described must be examined by your veterinarian to reveal the extent and cause of the problem and the appropriate therapy.

Sagging Eyes

Sometimes, llama owners may notice that one of their llamas has sagging lower eyelids which produce a haggard, worried look, rather

60

like a Basset Hound's eyes. Other llamas may be teary-eyed for no apparent reason. Often this is just an odd behavioral problem, caused by the llama being upset, tired, or apprehensive in some way. The key to figuring out the cause of the problem is to observe the llama when it is comfortable in its own familiar pasture or domain. If its eyes are normal then, and the phenomenon is seen only when the llama is caught, or put into some distasteful or threatening situation, consider it just another llama wonder.

Troubleshooting Eye Problems

1. **Any General Injury**

2. **Eyelid Injury**
   A. Minor injury; edges are in line.
   B. Fresh wound; gaping edges are not aligned or are cosmetically unacceptable.
   C. Old wound (more than 24 hrs.).

3. **Conjunctival Injury**
   A. Small wound; not bleeding.

   B. Large, gaping. or bleeding wound.

4. **Conjunctivitis**
   A. Red and teary eyes; no obvious wounds.

5. **Eyeball (Globe) Injury**
   A. Corneal opacity or injury.

   B. Large or puncture wounds; substances adhered to the eyeball.

6. **No abnormalities seen, but animal is squinting, teary, or holding eye closed.**

7. **Opacity noted inside the eye, visual difficulty noted.**

8. **Any signs of blindness**
   (Bumping into objects, picking up front legs excessively high, confused, with excessive ear and head turning)

---

1. Flush with sterile saline to determine location and extent of injury. Consult with your veterinarian. Check on tetanus vaccination status.

2.
   A. Consult with your veterinarian.
   B. Call veterinarian to suture defect.

   C. Consult with veterinarian for cosmetic surgery if it is cosmetically unacceptable or the eyelid rubs the eyeball.

3.
   A. Check eye for foreign body; consult with your veterinarian for appropriate therapy.
   B. Call veterinarian for possible suturing. If bleeding profusely, hold a gauze pack over the closed eye until bleeding slows.

4.
   A. Flush with sterile saline and check for foreign body. Call veterinarian for examination and treatment advice. Alert veterinarian if no improvement is seen after three days, or if the conjunctivitis is noticed in several of your llamas.

5.
   A. Have your veterinarian thoroughly examine eye(s) and recommend best therapy.
   B. Potentially serious; call veterinarian immediately. Don't touch the eye until the veterinarian arrives.

6. Call veterinarian.

7. Call veterinarian as the opacity may be in the lens of the eye (cataract) and may cause partial or total blindness in the affected eye.

8. Talk to your llama with a soothing voice. Place llama in an obstacle-free area until help arrives. Call your veterinarian for a complete ophthalmologic and neurologic work-up. If your llamas and alpacas are housed with horses, consider Equine herpes virus as a possible, though unlikely, cause of blindness.

CHAPTER 8:
WOUNDS

Introduction

Due to the llama's gentle temperament and high intelligence, wounds are not as common as in many other species. Occasionally, they do occur and must be managed successfully so that healing will be as quick and complete as possible. Wounds are classified as open, closed, or burns. Open wounds are lacerations or punctures where the entire thickness of the skin has been cut. The skin may be pulled apart and you may see the pinkish colored muscle and fatty tissues below the skin. If large blood vessels have been broken, the wound may bleed profusely. Closed wounds such as abrasions and bruises are those where the skin has not been broken all the way through.[1] Burns are caused by skin exposure to excessive heat, cold, or harmful corrosive chemicals.

Open Wound Care

On the trail, you may witness your llama's injury, but at home you may just discover a bloody area on your llama indicative of a wound. First, emergency care must be taken to stop any profuse hemorrhaging or bleeding. If the bleeding is excessive (steady drip or more), apply direct pressure to the wound for 10-20 minutes. If the bleeding wound is on a leg, apply a sterile pressure bandage (NOT A TOURNIQUET!) directly on to the wound to provide sufficient direct pressure to stop the bleeding. Most wounds will stop bleeding in about 20 minutes. If the bleeding was severe, do not remove the bandage until your llama can be seen by a veterinarian. Removal of the bandage may restart the bleeding.

If the injury is so severe that the blood loss is great, the animal may go into shock. Shock is a natural mechanism for animals to deal with severe injuries, infections, or excessive fluid loss. In order to try to stay alive, the heart rapidly pumps blood primarily to the vital organs such as the brain. Therefore, less vital organs and peripheral tissues do not receive adequate blood supply, and thus will be cool to the touch. In general, animals in shock will be weak, breathe rapidly, and have pale color of the mucous membranes and the skin. Shock itself may be life-threatening, even if the initial injury is not. An animal in shock needs IMMEDIATE shock treatment. This should be done by trained professionals. However, if your animal is in a state of shock, and weak, and no veterinarian is readily available, you must begin lifesaving shock therapy. First, stop any bleeding. Then, keep the animal warm and quiet. Offer it water or preferably water with electrolytes such as Lifeguard® or Gatorade® to drink. Offer small amounts frequently. This is less desirable than the IV treatment your veterinarian can administer, because when in a state of shock, the llama cannot absorb fluids efficiently through the gut. However, it may be the only choice when faced with a life or death emergency.

If hemorrhage and shock are not a problem, small amounts of fresh or dried blood can be gently cleaned from the wound to enable examination. Sterile saline is preferable, but if not available, use clean water. Always clean a wound from the wound to the periphery, so as not to drag hair, dirt and debris from the periphery into the wound.

Figure 8.1 BANDAGING

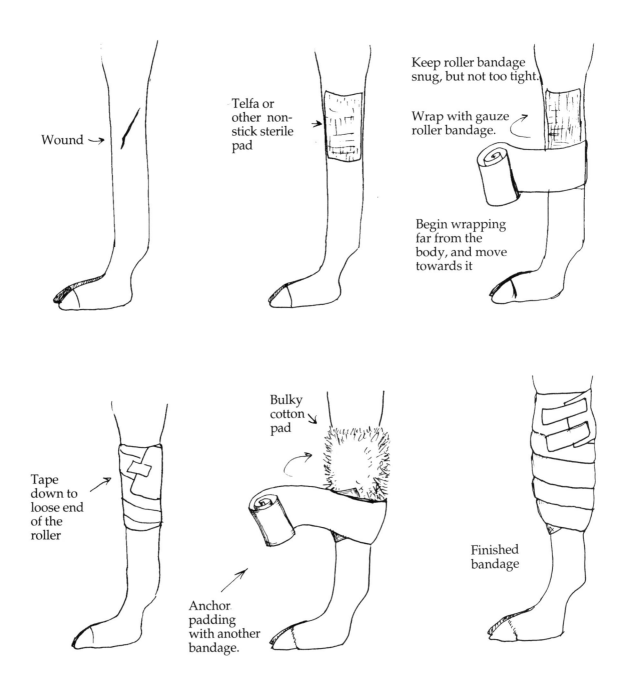

Wound →

Telfa or other non-stick sterile pad

Keep roller bandage snug, but not too tight.

Wrap with gauze roller bandage.

Begin wrapping far from the body, and move towards it

Tape down to loose end of the roller

Bulky cotton pad

Anchor padding with another bandage.

Finished bandage

If the wound is large or the skin edges are gaping apart, it's likely to need suturing by your veterinarian. Do not put antiseptics, wound ointments, or powders into wounds that may require suturing, as it makes cleaning the wound more difficult for the veterinarian. Also, some wound care products may even be quite irritating to the llama's tissues and cause pain and slow healing. If you suspect that the wound needs suturing, try to keep it clean until it can be sutured. If it is on a limb, apply a bandage such as clean gauze or a clean cloth and cotton. A diaper works well too, when held on with gauze wrapping. Bandages must be snug enough to stay on, but not so tight as to cut off circulation to the leg. After bandaging is complete, make sure you can slip a finger under the bandage.

In addition, make sure to check your records as to the date of the llama's last tetanus vaccination. If you are in an area where other Clostridial diseases are a problem, the llama may require these booster shots as well as for tetanus.

If the wound is minor, or an older wound that cannot be sutured due to contamination and retraction (drawing back) of the edges, it should still be examined by your veterinarian. Some wounds may be more serious than they appear. For example, wounds near joints must be treated aggressively to treat or prevent joint involvement. Your veterinarian may prescribe systemic antibiotics (oral, injectable) for wounds near joints, infected wounds, or those likely to become infected due to considerable contamination. Your veterinarian may prescribe anti-swelling and/or pain relief medication as deemed necessary. However, some leg wounds appear to be more painful than they truly are due to llamas' normal sensitivity to touch on their legs.

If upon examination, by or consultation with your veterinarian, it is decided that you can help with treatments yourself, or you are in a situation where veterinary assistance is not readily available, some general guidelines should include cleaning the edges of the wound with saline and Betadine® solution or scrub. Be careful not to carry any debris into the wound. Next, the hair should be clipped from the wound perimeter. Now, gently clean the wound itself. A general rule of thumb is never put anything into a wound that you wouldn't put into your own eye. In other words, be extremely gentle! Betadine® solution and saline solution work well to clean most wounds. Carefully remove all debris, hairs, and dirt from the wound itself. If the wound is extremely dirty, a Water Pik®, or a large needleless syringe will work well for flushing out dirt. After the wound has been thoroughly cleansed, a mild topical medication such as nitrofurazone ointment or powder may be applied.

If the wound is on the leg, in an area that is prone to getting dirty, a bandage should be applied. Bandages on the legs should be changed whenever they become moist or dirty. This initially may be daily for large wounds, but as healing progresses, may be every three to five days. If it is fly season, a commercial fly repellent may be used AROUND the wound or a screw worm spray could be used around the wound or directly on superficial wounds. Consult with your veterinarian for individualized treatment instructions specifically for your llama.

A common cause of open wounds in male llamas is the fighting that occurs between energetic males. When you find blood on a llama, check for a wound in the area. If none is found, the blood may have come from the other llama that may have been involved in a fight. If the male's long and sharp fighting teeth are not removed (see Chapter 6), they can be very damaging weapons. When males are fighting, they commonly bite each other on the neck, ears, legs, or near the testicles. These open wounds can be quite severe and often require stitches. Such wounds are also very contaminated with microorganisms, and are likely to abscess. The ear wounds may result in torn or shredded ears, which can lead to permanent holes and splits. These torn ears should be sutured for best cosmetic results. The structural portion of the ear is cartilage. In general, cartilage has poor blood supply and poor healing capabilities. The cartilage and the skin need to be sutured when the wound extends through the full thickness of the ear. Even then, some do not hold. However, without suturing, a split ear will remain split.

Puncture Wound Care

Puncture wounds can be dangerous, as they set up a perfect environment for anaerobic (oxygen disliking) bacteria such as the tetanus and the malignant edema organisms, and should always be examined by your veterinarian. These wounds should NOT be sutured, as they need to heal from the inside out, with the skin being the very last to close.

If, upon consultation with your veterinarian, you decide to treat a minor shallow wound yourself, a general method of therapy is to clean the wound periphery and clip the surrounding hairs as previously described. Next, check the wound for foreign material. Often a thorn or a stick may still be in the puncture and needs to be removed. The wound should then be flushed with tea-colored Betadine® and sterile saline solution. The wound should continue to be flushed daily until it has closed from the inside out, and all drainage has ceased. The llama should also receive boosters on its Clostridial vaccinations. Antibiotics are usually deemed necessary by your veterinarian depending on the depth and location of the puncture.

A non-healing puncture wound is often indicative of a foreign body. Such a body must be removed for proper healing. Any increase of swelling around the wound, foul odor, or crackly-feeling skin around the wound should receive immediate medical attention.

Abrasion Care

An abrasion or scrape does not go all the way through the skin. It only scrapes away the surface layers. After consultation with your veterinarian, minor wounds may be gently cleaned with a disinfectant. Debris, dirt, and hair should be removed from the wound. Antibiotic ointment should be applied and continued daily for several days. The abrasion will form a scab and then heal. If the scab is large, ointments should be used to keep the scab supple so that it won't crack.

Bruise Care

Bruises are usually minor. If caught immediately, cold packs should be held on the area to minimize the swelling. If the damage below the skin is extensive, hemorrhaging may occur and a fluid filled swelling will appear. These are pockets of blood or serum (the liquid portion of blood). These may need to be drained by your veterinarian if they do not disappear in a day or two.

Burn Care

Burns can be extremely serious. The skin's normal function is to keep bacteria out and help keep fluids in. Severe extensive burns lead to rapid dehydration. The llama must receive emergency fluid replacement therapy. Burns involving more than 15% of the animal's body need intensive fluid therapy, and it should be rushed to the nearest clinic. Burns are also very prone to infection. Your veterinarian will likely control the infection locally with antibiotic ointments, and systemically with injectable antibiotics. Burns are very painful, and pain relief medication is usually prescribed. Once the llama is stabilized, the burn wounds themselves are treated as open wounds. Follow-up care that may be done by you under your veterinarian's instructions may include topical ointments such as Betadine®. Again, tetanus boosters should be given as above.

Troubleshooting Wound Care

1. **Any wound**

2. **Bleeding wounds**

3. **Bleeding wound; Llama in shock**

4. **Fresh wound with skin edges gaping apart**

5. **Old wound** (more than 24 hours), **Contaminated wound**

6. **Puncture wound**

7. **Abrasion**

8. **Bruise**

9. **Burn**

---

1. Check vaccination status and give boosters if necessary.

2. Stop the bleeding. Apply direct pressure until bleeding stops, or apply a pressure bandage and call your veterinarian.

3. Stop bleeding as above. Keep llama warm and quiet. Seek veterinary care A.S.A.P.

4. Probably needs suturing. Keep wound clean or protected until it can be seen by a veterinarian. Do not apply wound dressings.

5. Call veterinarian. Minor wound care may be done by you under veterinary supervision. Your veterinarian may prescribe systemic antibiotics depending on location and contamination.

6. Call veterinarian. Minor puncture wound care may be done by you under veterinary supervision. Generally these wounds are not sutured as they must heal from the inside out. Your veterinarian may prescribe systemic antibiotics.

7. Consult with your veterinarian. Clean and apply antibiotic ointment.

8. Cold packs initially. Call veterinarian for drainage if a fluid pocket develops and persists.

9. If more than 15% of the body is burned, it is an EMERGENCY. Llama needs immediate shock and fluid therapy from your veterinarian. Otherwise, treat as directed by your veterinarian.

NOTES

CHAPTER 9:
LUMPS AND BUMPS

Introduction

When handling and grooming your llama and alpaca, you should always be aware of any changes of the skin or coat. Occasionally, you find a lump or bump on your animal. It may have suddenly appeared, or you may not have previously noticed it. It may be a normal bump. First, check the same place on the other side of the animal. If there is an identical bump on the other side or other leg, it is probably a part of your animal's normal anatomy. If no matching lump is found, it may indicate a problem.

Scars

The discovered lump may be a result of an injury. Does the llama have a current injury, or did it previously have an injury at the site of the lump? If so, it could be scar tissue as result of the normal healing process. Scar tissue is often thick and grey appearing. Touching the scar tissue is generally not painful to the llama. If the lump is painful to the llama, it is indicative of a more serious or more recent problem. For example, a rock-hard lump may form around a bone chip or a wooden splinter in the tissues. These wounds are generally painful, oozing liquid, and often form an abscess.

Swelling from Wounds

Shortly after a wound occurs, swelling begins. Swelling will subside if the wound is treated appropriately. In general, cold packs applied immediately following an injury will minimize swelling. About 24 hours after the injury occurs, hot packs will increase blood supply to carry away swelling. Fifteen to twenty minutes of holding the packs on the wound twice daily is sufficient.

Blunt trauma such as running into the fence or a kick may cause swelling even if the skin is not broken. The swelling should be treated as above. Sometimes, the only evidence of the injury may be a large fluid filled swelling. Often a blood vessel is ruptured inside, under the skin, resulting in blood or serum accumulation. The swelling is soft and the fluid can be felt inside. These are called hematomas when they are filled with blood, and seromas when filled with serum. With hot packs or soaks, they may resolve themselves. If not, they must be drained by your veterinarian.

Allergies

Allergies are caused by the body's immune or defense system reacting to certain substances. Usually the allergic substances or allergens are foreign substances, but occasionally, the body's immune system may even attack its own tissues. Common allergens are pollen, weeds, certain feeds, drugs, parasites, insect bites or stings, and insecticides. These substances cause cells in the llama's body to release chemicals such as histamine. These chemicals in turn cause the signs observed with allergies. The signs may range from a localized swelling to respiratory difficulty and shock.

With a localized allergic reaction, chemical mediators released cause a swelling immediately around the allergen such as the bug bite or injection site. It may be only a localized irritation or a true allergic response. The swelling is often very minor and quickly disappears without therapy. In more serious reactions, the swelling may continue to increase until it could cause some significant damage. For example, an insect sting on the face or throat may cause enough swelling to obstruct the airways and hamper breathing. These cases need aggressive veterinary therapy which includes anti-swelling medications. Later, antibiotics may be needed to control secondary infection.

Allergic reactions may also cause your llama to "break out" in bumps or wheals all over its body. The wheals are soft and short-lived, provided that the offending cause is removed. Often, however, you cannot figure out what caused the allergy. In these cases, the llama should be taken off of pasture and kept in a corral so that you can control its environment and feeding. If the allergens were from the pasture, the llama's bumps will disappear within a day or two. Be cautious when administering medications, new feeds, or insecticides, as they too may have been the inciting allergen. If you still don't know the offending allergen by the time the bumps have disappeared, use extreme caution and careful observation when reintroducing drugs and insecticides. A second allergic reaction is often more serious than the first.

Severe allergic reactions may be life-threatening. The chemical mediators may cause narrowing of the airways, making breathing difficult. Mucus is often secreted into the airways, which further impedes respiration. The chemicals released also may cause some blood vessels to become more permeable or leaky. Fluids are then lost from the blood vessels and shock occurs. Thus, if the allergic reaction is severe or is accompanied by other signs of distress such as respira-

tory difficulty, weakness, hemorrhaging, or convulsions, the llama needs immediate medical therapy by your veterinarian. Always consider an allergic reaction whenever the llama has any severe reactions as above, or has one or many bumps that suddenly appear. The recent history of the llama may also help indicate an allergic reaction, especially if you have just medicated your llama, changed feed, changed pastures, or if insects are numerous.

Tumors

Llamas can also develop tumors of the skin or tumors below the skin. These are usually slow growing lumps that do not get smaller. Some are covered with skin, while some are ulcerated sores that will not heal. If a sore or lump does not heal or continues to grow in size, have your veterinarian examine it. If a tumor is diagnosed, the type and general prognosis can be determined. Some tumors are benign or innocent tumors such as fatty tumors. These tend to remain localized and do not spread throughout the body. Others may be quite invasive and malignant. Once the type and tumor behavior are determined, appropriate therapy may begin.

Abscesses

Finally, llamas seem to be prone to abscesses. An abscess is basically a pocket of pus. It occurs as a result of a wound, fighting tooth injury, an infected lymph node, an infection in the bloodstream, or an imbedded foreign body such as a foxtail or splinter. Sometimes the inciting cause cannot be found. Commonly, a lump is suddenly noticed on the llama. The lump is firm to slightly soft, warm to the touch, and painful to the llama. The abscess will need to be surgically removed, if possible, or lanced and drained to remove the pus by your veterinarian. However, it cannot be lanced until it is softened and pointed (comes to a "head"). The abscess can be softened by hot soaks for 20 minutes twice to three times daily. A washcloth dipped in hot water, a hot water bottle, reusable hot pack, or hot water in a Ziploc ®bag works well. Be careful that the water is not so hot that it burns your llama. Hold the cloth over the abscess until the cloth cools. Then re-dip the cloth in the hot water and repeat until you've endured the 20 minutes. A heating pad will work too; provided it is not set too hot as to burn the llama's skin. The heating pad will get quite messy, though, once the abscess has opened.

The abscess will occasionally rupture on its own with the heat therapy only. Once the abscess is open and draining, it needs daily care so that it won't reform. Any pus should be cleaned from the abscess daily. Gently squeezing the abscess often helps to get the pus out. Continued hot soaks will keep "drawing" the pus out through the opening. Adding a sprinkle of Epson Salts to your soaking cloth provides some additional "drawing" ability. The drainage hole of the abscess must be kept open or else the abscess will reform. The abscess needs to heal from the inside out, so that the surface of the skin is the LAST to heal.

Antibiotics are often used in treating abscesses. A culture and sensitivity test may be done by your veterinarian. A culture involves carefully taking a sample or swab from the abscess and growing it in the laboratory to determine the offending microorganism. Once the microorganism is found, a sensitivity test is run. This involves testing various antibiotics to find out which is most effective in combating the infection. Abscesses caused by *Actinomyces* bacteria,

which often occur on the animals' head and around tooth roots, can be particularly slow to heal. Treatment, including antibiotics, often takes 1 to 1-1/2 months.[1]

Daily flushing of the abscess hole with an antiseptic such as Betadine® solution may be used to help combat the infection. A syringe with a curved blunt tip works well for the flushing. If the abscess is in a sensitive area, such as around the eyes, the Betadine® should be diluted with sterile saline solution to a tea color. Flush the abscess repeatedly until no more pus comes out and only the solution comes back out. If desired, some antiseptic ointment can be put over the abscess when you are finished with your daily treatment. Care should be taken to clean the pus from the fiber, the surroundings, and the halter, as sometimes the bacteria in the pus are contagious to people or other animals. Thus, you need to use common sense and basic cleanliness, such as washing your hands, to prevent spreading of the infection. The pus is also very irritating to the adjacent normal skin. A little bit of Vaseline® around the abscess will prevent the pus from sticking to or irritating the skin. If the abscess does not heal or spreads to other sites, it should be reexamined by your veterinarian. Often a foreign body remains in the abscess and needs to be removed. The average uncomplicated abscess should heal in about two weeks. Obviously, this varies with the size of the abscess.

NOTES

CHAPTER 10:
SKIN DISORDERS

Introduction

The llama's skin is made of two layers. The outer layer is called the epidermis. The epidermis serves as a wrapper, a barrier against microorganisms, and prevents water loss and absorption. The lower layer is the dermis. It contains the blood vessels and nerves. It supports the epidermis. Sweat glands are found in llamas and, just as in people, help the llamas to cool themselves by evaporative cooling. They are located most abundantly and effectively on the underside of the body. Llamas have two fiber (hair) types. The primary fibers are the guard fibers. These are the large coarse fibers. The secondary fibers are the finer undercoat fibers. This undercoat accounts for the fineness and softness of alpaca fiber in which guard hairs are very few or absent.

Hair Loss

Llamas periodically lose hairs as new hairs emerge from the hair follicles and the old ones are forced out, or as the hair fiber may weaken and break. Most patterns of shedding do not cause actual bald spots, since not all hairs are shed in a particular area at the same time and new hair growth is rapidly emerging from the skin. Alopecia is the term used for complete bald spots with no new hairs emerging from the skin. Although it is hypothesized that some localized areas of alopecia may be "normal" shedding patterns, any persistent or spread-ing bald spot should be examined by your veterinarian, especially if accompanied by flaky, reddened, moist, or roughened skin. One common spot of alopecia is on the bridge of the nose. It seems to be more common in dark faced animals and more common the summer. Causes include halter rubbing, fly irritation and self-trauma. Usually the hairs grow back over the winter.

Ringworm will also manifest as spots of hair loss. Ringworm is caused by a fungus that grows in the hair and hair follicle. The infected hair breaks or falls out, thus causing a spot of hair loss. The spot is usually scaly and crusty. The disease is contagious to other animals and people. Diagnosis may be made by a culture or by microscopic examination of the hairs for fungal spores. Treatment includes topical Betadine®, thiabendazole, or captan (3%). Occasionally your veterinarian must administer systemic antifungal therapy.

Bald spots and lack of fiber luster and growth are also indicative of overall health of the animal. Poor nutrition, parasitism and many systemic illnesses manifest in poor skin and fiber.

Mange

Of the diseases that affect the llama's skin, mange is one of the most uncomfortable for the llama. Mange is caused by mite infestation of the skin. The mite of one of several species buries itself in the epidermis. Initially, it causes small raised red bumps and then leads to skin crusting and thickening. It is often very itchy to the llama. This leads to scratching and self-trauma, which causes very thickened and darkly pigmented skin. Most lesions or sores are on the short fiber areas such as the belly, chest, and legs. The mite can be diagnosed microscopically by your veterinarian, with a skin scraping. Treatment begins with isolation, as mange is contagious. The mite can be killed with Ivomec® injected SQ. Usually only one treatment is needed. However, a second injection in about three weeks may help in severe cases. (See Chapter 5 for additional information on mange.)

Lice

Another parasite affecting llama skin is the louse (lice pl.). Lice prefer to live in the heavily fibered areas such as the back and the hip region. They cause a dull look to the fiber and many broken hairs. The skin has excessive dandruff where infested. The lice often cause the skin to itch. The lice or their eggs (nits) can be seen on an infested llama. Treatment includes Ivomec® for sucking lice, Carbaryl or Methoxychlor[1] powder applied thoroughly to the skin, or pour-ons such as Tiguvon®. (See Chapter 5 for further information on lice.)

Allergies

Allergies may also manifest as skin disease. These are seasonal problems in cases of pollens and weeds. Non-seasonal allergies could be to food, dust, or molds. Signs can be excessive itching, hair loss, increased skin thickness, red sores, and lumps. The diagnosis is based on elimination of the other causes of similar skin lesions. A skin biopsy done by your veterinarian is helpful in diagnosis, too. Treatment ideally is to remove the offending substance. If this is not possible, anti-allergy medication and desensitization are alternatives.

Zinc Responsive Skin Disease

Another llama condition is a zinc responsive skin disease. At present, the cause is unknown, but zinc supplementation seems to help in therapy of this disease. Areas of hair loss appear and the skin becomes very thickened, scaly, and heavily pigmented, resembling elephant skin. It is usually not itchy. It tends to involve the face, inner thighs, and the underneath part of the belly. Confirmation of the disease is based on a negative skin scraping, a skin biopsy, and response to therapy. Don't forget that mange has very similar skin changes. Remission following zinc supplementation begins in about two months. The improvement is slow and may take up to a year for significant improvement. To maintain remission, continued zinc supplementation is probably required. Consult with your veterinarian as to the best way to supplement zinc if it is needed.[2]

Valley Fever

Valley Fever (Coccidioidomycosis) is caused by a fungus, *Coccidioides immitis*, found in hot and dry areas of the southwestern United States. Many species, including humans, dogs, llamas, and alpacas, may become infected by inhaling fungal spores from contaminated soil and dust particles. The disease, however, is not contagious between animals. Signs of the infection vary, but include open or raised sores on the skin, chronic cough, lameness, neurologic abnormalities, fever, and lack of appetite. Your veterinarian may suspect Valley Fever if your animal has any of these signs and you are in an area known to have this fungus in the soil or have recently traveled to the Southwest with your animal. A diagnosis is often difficult, but the fungus (in various stages) may be identified from infected tissues (such as the open sores) or from discharge from infected respiratory structures. The disease is frequently too advanced to treat upon diagnosis. Anti-fungal medicines, although very costly, have offered improvement in other species, but have had poor results in llamas.[3]

CHAPTER 11: *LAMENESS*

Introduction

Llamas and alpacas normally walk on their middle and distal (furthest out) bones in their feet, called phalanges. These are anatomically equivalent to the bones of the middle part and tip of our fingers. The toenail surrounds and protrudes from the tip of each distal phalanx. The pad on the bottom of the foot resembles a dog's foot pad, but is much tougher.

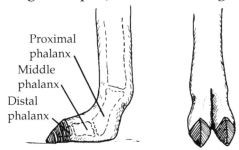

Proximal
phalanx
Middle
phalanx
Distal
phalanx

Figure 11.1 **NORMAL FEET**

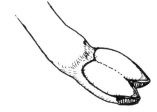

Figure 11.2 **NORMAL FOOT PAD**

Lameness or limping is an indication of pain or mechanical problems in one or more of a llama's legs. Among the most common causes of lameness are foot problems; either nails that are too long or damaged, or bruises, cracks, cuts, or infection of the foot pad. Other possible causes include sagging or dropped ankles, sprains and injuries higher up the leg.

Toenail Trimming

Llamas' toenails grow continuously throughout their lives. Llamas that live on hard, rocky ground, or who have plenty of exercise on the road or trail naturally wear their nails short. Llamas that are sedentary, have little space in which to exercise, or live on soft ground, will tend to grow long toe-

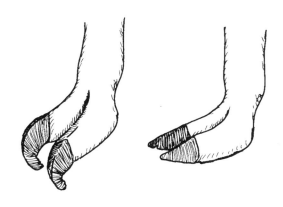

Figure 11.3 **LONG TOENAILS**

nails. These long toenails may curve, cause the toe to twist, pinch the pad, or break off painfully. Toenail problems like these may cause lameness, and for llamas with such problems, toenail trimming should be a part of routine care.

75

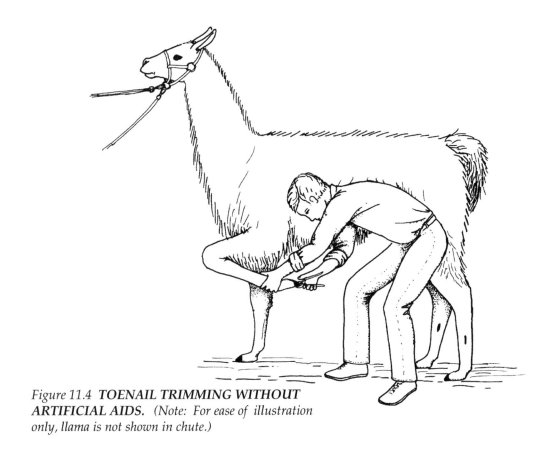

Figure 11.4 **TOENAIL TRIMMING WITHOUT ARTIFICIAL AIDS.** *(Note: For ease of illustration only, llama is not shown in chute.)*

Equipment:

While some llamas are so well desensitized that they will allow you to pick up their feet and trim their nails almost anywhere, restraint will be necessary for most llamas. Nails can be trimmed in almost any kind of chute, but the techniques will vary. In large, closed sided chutes, the llama, nail trimmer and foot holder will all be inside the chute together, while in smaller or open sided chutes the nail trimmer and holder stay outside, and move around the llama tied inside.

Hoof or nail nippers are essential, and can be purchased at ranch, horse or horseshoeing supply stores. Horse hoof nippers work well, but can be large and unwieldy, requiring two hands to use. Horseshoe nail nippers are excellent, as long as they haven't yet been used to cut horseshoe nails! Electricians' end nippers work fairly well, but are not quite as sharp. Also, due to their construction, it may be difficult to cut as close to the edge of the nail as desired. Some people have used small scissor-type pruning shears, foot rot shears, or even tin snips, and been able to do the job. One advantage of spring loaded scissor-type

nippers is that they can be used easily with one hand.

Procedure:

Nail trimming is usually easier when the llama's feet are wet and the nails soft. Therefore, if you have a choice, bad weather is a good time to trim nails. It is quite possible for one person to trim a llama's nails, but it is much easier, especially when starting out, to have two people. One person will hold the llama's foot in the proper position, either by hand or with the stake line, as described below, while the other will do the actual trimming. Begin with the llama cross-tied in the chute, or the stanchion, as is appropriate for your chute. With the holder picking up each foot and keeping it in the right position for trimming, move around the llama, trimming each foot before going on to the next. Some llamas show their objection to the procedure by kushing. It is possible to trim toenails while the llama is in kush position by pulling the foot out from underneath the llama's body. However, llamas are amazingly able to hide their feet directly under the heaviest part of their body! Sometimes it is possible to trim a

76

calm female llama's front toenails while she is being bred, if she rolls on her side. Be creative, and look for occasions that might be easily adapted to your needs.

An excellent method, developed by Bobra B. Goldsmith, uses a twenty foot stake or lunge line as a tool to hold the llama's foot off the ground and in the proper position. The line should be made of flat woven webbing, and have a small, hand-sized loop sewn in one end. An open sided chute is required to use this method.

Bobra Goldsmith's method works especially well for llamas that have not been completely desensitized, as well as for well-trained llamas. It is also valuable when the llama's

handlers are not especially large or strong. With the llama cross-tied in an open sided chute, pass the loop end of the twenty foot stake line around the leg you will be working on. Some llamas will be quite calm about this; others will be highly offended. If the llama is hard to work with, you can either try to pass the loop quickly around the leg, high up and close to the body, or you can place the stake line on the ground, and maneuver the llama until it steps into the right position. Next, pick up the sewn loop, pass a fold or bight of the webbing through it, and pull on the loose end of the webbing until you have an unsewn loop of webbing about five to seven feet long.

a.

Figure 11.5 (a-d) PICKING UP FEET FOR TOENAIL TRIMMING.

b.

c.

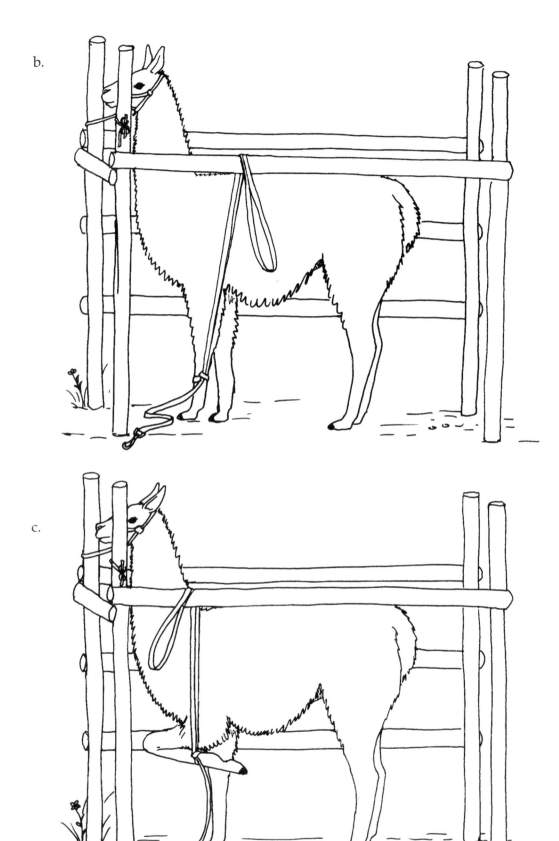

d.

A slip knot forms around the llama's leg, created out of the sewn loop and the first few inches of webbing. Pass this long, unsewn loop up the inside of the chute, and over the top rail. Leave the loose end with the clip dangling unless it is in the way. It can be held gently with the webbing of the loop, but pulling on it will loosen the slip knot around the llama's leg. When the long loop is pulled downward, the llama's leg will be lifted off the ground. When it is at the right height, wrap the loop two or three times around the rail, and hold it while the trimmer takes care of the nail. The mechanical advantage produced by wrapping the stake line around the rail makes it easy to keep the foot up for as long as it takes to trim the nail. If the llama paws wildly during this procedure, let the leg down briefly, reassure the llama, and try again. Most llamas become accustomed to the process very

quickly, but for the rare llama that tries to lie down, cinches can be used to keep it standing.

When trimming the front toenails, the slip knot should be located at about the middle of the lower leg. For the back toenails, the slip knot should be just below the hock. The holder can easily keep the leg in the correct position for some time by wrapping the end of the loop several times around the rail. If one person is doing the entire trimming process, the end of the loop can be tied to the rail with a quick release knot.

Trimming toenails is not particularly difficult. If the llama has been on soft or muddy ground, you will need to use the nail nipper points, a hoof pick, or even a bit of wood to remove the mud or manure so that you can see the pad, quick, and nail, and can tell where you need to trim.

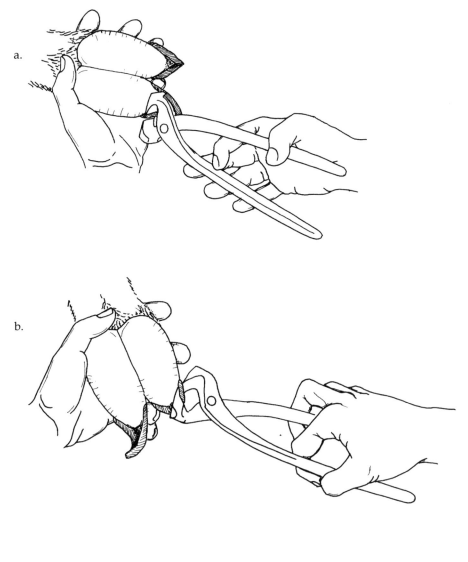

a.

b.

c.

Cut 1
Cut 2
Cut 6
Cut 3
Cut 4   Cut 5

#1 Be cautious here; if the nail is folded over, pressing on the pad, you may need to trim around the folded area. Then, trim the nail later, when it is wet and more flexible.
#2 This area is the "quick". Don't cut too close to it as it will hurt and bleed if damaged.

1.   2.

Figure 11.6 **TRIMMING TOENAILS**

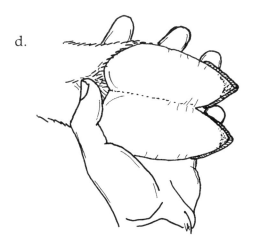

d.

d. Finished toenail.

Start at the back edges of each nail, and work forward towards the point. Don't cut any closer to the edge of the pad or quick (the soft tissue just behind the point of the nail) than 1/16 or even 1/8 inch, since cuts here will bleed, be painful, and expose the llama to risk of infection. It is probably a good idea for a beginner to be conservative at first and leave a larger margin for error. With more experience it is easier to trim the nails short without hurting the llama. When trimming the insides of the toenails, it helps to use your fingers to hold the toes apart and provide more space for the nail nippers. Once both sides of a nail have been trimmed, it is time to remove the point with one final cut perpendicular to the line of the nail. Don't cut too close to the quick, and avoid pinching the llama's pad between the two sides of the nail.

Special Problems:

Long, curved nails may have caused the llama's entire toe to twist uncomfortably. One trimming may not be enough to allow the toe and nail to grow straight. If this is a problem for a particular llama, it is important to trim that llama's nails regularly; at least once per month, until the nails and toes remain straight.

When the edge of a toenail has folded over to press on the pad, don't try to simply unfold it and cut if off. Trim the surrounding nail, then try to file the edge gently with a fine-toothed file. If this is not successful, the folded area may simply wear off once the adjacent long nails have been removed, or you can try to trim it off later when the nail is damp and somewhat more flexible.

If you do accidentally cut into the quick and cause some bleeding, go ahead and finish trimming the nail. When you are done trimming, if the bleeding hasn't stopped, apply direct pressure until it does. Then put on some 7% iodine which will disinfect and stop the bleeding. Watch the llama over the next few days in case infection or lameness develops.

Toenails can also be torn off, or be injured and gradually fall off. Treatment includes general open wound therapy. (See Chapter 8.) The nail bed may need to be protected from the mud and dirt with a bandage or bootee. If the foot is kept free of infection, the toenails will usually grow back without further care.

81

Pad Problems

Problems with the pad of the foot occur infrequently but are the most common cause of lameness. When you examine the llama's pad, you may see cracks, cuts, thin and worn looking areas, bumps, blisters, raw and infected sores, or even foreign bodies such as nails, glass or thorns. Prolonged and excessive exposure to a wet environment could also result in pad problems. These injuries should be treated soon to prevent them from getting worse. Foreign objects should be removed, and any fresh cuts should be sutured to speed up healing. Smaller or older cuts or sores need to be cleaned and then protected. A good method of treating these problems is to wipe off the superficial dirt, and then have the llama stand in a shallow container such as a rubber feed tub. Fill the container with several inches (depending on the location of the injury) of warm water, a handful of Epsom Salts, and a bit of Betadine® solution. You may need to steady the container at first to keep the llama from overturning it, but llamas usually find the procedure soothing and seem to enjoy it. For minor injuries, this treatment should be repeated daily until the sores heal. Deeper, more severe injuries need protection from dirt and irritation after the foot soak. Daily bandaging can be costly and may be challenging to put on a wiggly, uncooperative llama. Bootees can be improvised from nylon jacket material, socks, moccasins, or leather, and tied, laced or taped on. Make sure that the fastening is not too tight. Your veterinarian may suggest antibiotic usage if the sores are infected, and analgesics if the llama is in pain. Additionally, if the animal's environment is contributing to pad problems, improve housing or pasture conditions.

Abscesses can also occur in the pad, and will make the llama extremely lame. Sometimes the pain is so severe that the llama will refuse to put any weight on the foot. Such an abscess may break open and start to drain on or above the pad, where the hair begins, as the skin is not as tough there. Care includes soaking the foot as described above, to draw out the infection, plus antibiotic therapy as recommended by your veterinarian.

Joint and Leg Problems

Some llamas have what appears to be sagging or dropped ankles. This abnormality may cause varying degrees of lameness.

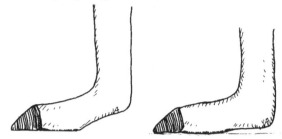

Figure ll.7 DROPPED ANKLES

The ankle or fetlock is the joint between the metacarpus or metatarsus (front or back cannon bone) and the first phalanx or upper part of the foot (see Figure 11.1). Normally the joint is stabilized by ligaments and tendons. When the joint loses stability or is malformed, the ankle may touch or nearly touch the ground. Dropped ankles are seen more commonly in the front legs because the front legs support more of the llama's weight than the back legs. Causes of dropped ankles are not completely known. In older llamas, years of wear and tear may bring about a gradual degeneration of the joint and subsequent loss of stability. Llamas who are very over-weight put more strain on the joints. Having a llama carry heavy loads before it is full grown will also tend to strain the joint. Other causes include congenital deformities or improper nutrition. If your llama is showing a tendency to having sagging ankles, do not further stress the joints by packing or exercising the llama until it has been seen by your veterinarian. Once the ankles are severely dropped, the change is permanent, and the llama will be lame due to the pain from bruising, arthritis, and excessive strain on the tendons and ligaments.

These complications can occur as a result of other injuries and/or abnormalities besides dropped ankles. Bruises, arthritis, sprains, muscle strains, or stretched tendons can occur anywhere along the leg. With these leg injuries, allow the llama to rest, and consult with your veterinarian. If the llama is very lame, does not improve with rest, or has a large area of swelling, veterinary attention is needed promptly.

Leg Fractures

Llamas are active animals whether they are working, playing, or fighting. Fractures can occur as abnormal stresses are put on a bone as a result of a stumble, twist, wrong step, or an encounter with a solid object. If your llama cannot put ANY weight on a leg, or has an obvious fracture, call your veterinarian for immediate assistance. Do not move the animal until the veterinarian arrives or provides further instructions. If the llama must be transported in order to receive medical attention, the leg should be splinted. To splint a leg properly, you must know where it is fractured. This may be obvious due to localized bleeding, swelling, or an abnormal bend from which the rest of the leg dangles. If you can tell where the fracture is, immobilize the leg, including the joint above the fracture and the joint below it. Upper leg fractures (humerus or femur) are not usually amenable to external bandage supports, and the animal should be transported as is to the hospital. Calm llamas may accept a sling securing a flexed front leg against the chest.

The key to good splinting is to use plenty of padding. Cotton, towels, pillows, or even torn sheets or diapers will work. Hold the padding on with gauze roller bandages, making sure to cover the fracture site and the joints on either side of it. Finally, apply a splint such as a length of PVC pipe cut in half, or a strip of plywood, but use any splint only OVER the padding. If no such splint is available, just apply several bulky layers of bandage material as support, but make sure the bandages are not too tight. If you cannot tell where the fracture or injury is, do not try to splint it, as you may cause further damage. Also, never attempt to realign fractured pieces on your own!

Llamas can move on only three legs, so proceed slowly and use common sense in moving the llama into a trailer. When traveling over rough terrain, either on foot or in a trailer, it may be desirable to use a sling or cinches to help support the llama's weight. Now the llama can be taken to a facility for proper diagnosis and appropriate therapy.

NOTES

CHAPTER 12:
HEAT AND COLD PROBLEMS

Introduction

Llamas and alpacas are native to the puna or altiplano of South America. This is an area of high altitude (12,000 to 14,000 feet, or more), but because it is near the equator, the temperatures tend to be warmer and show less seasonal variation than one might expect in the high mountains. Generally, temperatures vary between 20 and 60 degrees F (-5 and 15 degrees C), neither very hot nor extremely cold. Consequently, although llamas and alpacas are quite adaptable, they do have trouble coping with extremes of temperature, either very high or very low. Of the two, high temperatures, especially when associated with high humidity, seem to be most difficult to deal with successfully.

Hyperthermia

Hyperthermia or heat stroke occurs when the animal's core body temperature rises above normal. Death occurs when the body temperature rises 10 degrees above normal body temperature.[1] Heat stress is more serious in areas that have both heat and humidity, but can also occur as llamas exercise, fight, have a fever, are down for long periods of time (see Chapter 15), or from being exposed to excessive stagnated air.[2] Fat llamas definitely have an increased likelihood of hyperthermia than do the thin ones. Also, newborn babies do not seem to thrive as well when born in the heat of the summer. The babies easily become dehydrated and don't yet know how to avoid the hottest and sunniest part of their pen.

All animals tend to be lazier and more lethargic in hot weather. If the llama begins to get heat stress, the rectal temperature will begin to rise to 103 to 104 degrees F. As the llama begins to overheat, it will sweat to try to cool itself. If the llama does not get relief at this point, it will begin open-mouth panting to try to dissipate heat and may cease sweating due to dehydration. If the stress continues, the llama will begin to walk in an uncoordinated fashion, may seizure, will eventually go down and could die. Hyperthermia also adversely affects reproduction by decreasing male fertility and causing defects in unborn babies, some of which can be quite severe.

Obviously, hyperthermia is an emergency situation! Treatment first depends on recognizing a problem exists. If your llama is showing any of the above signs, or is just not acting right when it is hot outside, take a rectal temperature. If elevated, move the llama to shade or into a shelter immediately! A baby can even be immersed into cool water. Be careful not to drown the baby, though. Adults can be sprayed with water or sprayed with rubbing alcohol on the short-fibered areas to help bring the temperature down. If a fan is available, use it. Monitor the rectal temperatures. If the temperature does not come down, call your veterinarian as soon as possible.

If your llama shows signs of dehydration or water loss, offer it water or water with electrolytes, and consult with your veterinarian. Signs of dehydration include decreased skin elasticity (check this in short-fibered areas such as cheeks and chest), sunken eyes, dry mouth, absence of urination, and when the membranes in the mouth are blanched with digital pressure, the return of pink color is slower than two seconds, indicating delayed capillary refill time.

Once the llama recovers enough to stand and the rectal temperature is below 102 degrees F, keep it in a cool and shady area. A sprinkler can be used to aid in cooling. Don't allow an animal that has become overheated to go back to the hot environment, because it will have a relapse. Even when the llama is back to normal, keep a close eye on it for signs of heat stroke again, if the weather continues to be hot. This llama will be prone to reoccurrence.

Ingestion of toxic tall fescue grass may be related to hyperthermia in llamas too. Tall fescue grass is a hardy perennial grass that is common in central and southeastern United States. Fescue infected with a parasitic fungus can cause poisoning in cattle, sheep and horses. Signs of poisoning in these species include lameness, dry gangrene of one or more feet, ears and tail, dystocia, poor weight gain, decrease or lack of milk production, and hyperthermia. Although the effects of toxic fescue in llamas are not known, owners should avoid grazing their llamas in infected tall fescue pastures - especially during hot weather. Check with your veterinarian in regards to the risk of fescue poisoning in your area.[3]

Ideally, hyperthermia should be prevented. Tips on prevention include keeping fresh clean water available at ALL times for ALL llamas. Shade should always be available. Children's wading pools also offer relief from the heat, as the llamas can stand or lie in cool water. Make sure to keep the pool clean and in the shade. Sprinklers should be set up for hot days, too. The barn area should have a fan to avoid turning it into an oven. Also, keep your animal in good condition, because healthy animals are less prone to heat stress. This includes keeping your llamas from getting too fat.

In addition, stressful procedures such as vaccinating, deworming, and traveling should be done on cool days. If you stress an animal in the heat, it is more prone to hyperthermia. When obtaining a new animal, do so during the cooler seasons to prevent stress, and allow time for it to acclimatize before the hot weather begins.

Avoid transporting llamas in hot weather. If this in unavoidable, an air conditioned van is quite comfortable. Heat can build up quickly in a trailer, even if it is well ventilated. If environmental temperatures are above 90 degrees F, it is advisable to cool your llamas by spraying or even soaking their bodies with water periodically during the trip. In addition, provide ample opportunities for your llamas to drink water.

If you raise llamas or alpacas in a hot and humid climate, shear your heavy-fibered animals either partially or completely down to a couple of inches of fiber. Depending on the climate, alpacas should be sheared every year or two. Do not shear to the skin (1-3 inches is adequate), or else the skin will sunburn and the body temperature will rise. Shear them in advance of the hot weather, so that they can get used to the short fiber. (See Appendix VII for a resource list on fiber and shearing.) With common sense, such as planning for babies to be born in slightly cooler months, and the above preventative procedures, hopefully, you won't have to experience hyperthermia with your animals.

Hypothermia

Hypothermia is a fall in the internal body temperature due to disease or exposure to extreme cold temperatures. Hypothermia begins in most species when the body temperature drops below 90 degrees F. However, in babies, you should be concerned with a temperature below 100 degrees F; they don't have the fat insulation yet, their hair isn't as heavy, and they may also be wet if just born. The lower the temperature drops, the worse the prognosis. If hypothermia is associated with disease (e.g., pneumonia or peritonitis), it is a very poor prognostic sign.

Hypothermia occurs when the llama's normal protective thermal mechanisms are not enough to maintain a normal body temperature. Llamas have abundant fiber which provides excellent insulation. Llamas also have a fat layer under the skin for further insulation. When this insulation alone cannot keep the llama warm, it will start to shiver. Shivering is a normal reaction to cold environmental temperatures. It is caused by quick muscle tremors that produce heat. However, these muscle tremors require a lot of energy and oxygen. As energy reserves are depleted, shivering decreases and the body temperature falls. This causes the heart rate and breathing to slow, the blood pressure to drop, and eventually unconsciousness occurs.

A diagnosis of hypothermia is based on history of exposure and rectal temperature. However, standard rectal thermometers only measure as low as 93 degrees F and severe hypothermia may be missed. Animals, especially babies, with severe hypothermia often appear dead. They may be comatose with barely detectable vital signs. Moderate signs of hypothermia include lethargy, incoordination and muscle stiffness.

Hypothermia is a serious situation and your veterinarian should be contacted. Treatment involves general life support, preventing further loss of heat, and re-warming. The animal should be gently brought into a heated room and allowed to warm until its rectal temperature returns to normal. Additional methods of warming include blankets, heat lamps, hair dryers, warm water baths, hot water bottles, and providing warm water to drink.

In newborns, and rarely in older animals, a cold environmental temperature may not actually cause hypothermia, but may freeze the ear tips. When it is cold, the blood vessels in the ears constrict to prevent chilling of the circulating blood. This decreases circulation, causing further cooling of the ears since they are not receiving as much warm blood. Basically, the frozen tips die. Feeling is lost in the tips of the ears and eventually the tips dry and fall off.

Extremely cold environmental temperatures dictate hypothermia prevention for all llamas, although the young, short-fibered, thin, old and ill llamas are at a higher risk. The llamas should be provided shelter and observed closely as some may even need to be allowed into a heated room. Blankets, sweaters, and legs from long underwear for neck warmers help. However, never allow clothing or blankets to remain on the animal once they become wet. If possible, time the birthing season to avoid the coldest periods. When births occur in the cold, get the newborns in out of the cold, especially out of the wind. It is important to dry newborns (including their ears), as wetness increases the chance of freezing.

Prevention also includes a high level of nutrition. Water must be thawed to ensure adequate hydration. Roughage amounts (hay) can be increased as digestion of the roughage releases heat. In addition, feed energy levels should increase to compensate for the increased energy needs of keeping warm. A large amount of energy is used for shivering. Concentrates or grains are an excellent compact energy source. Corn is very high in usable energy. However, add grain to the llama's diet gradually to avoid digestive upsets.

NOTES

CHAPTER 13:
RESPIRATORY PROBLEMS

Introduction

Normal respiration, or breathing, in the llama or alpaca should be quiet and effortless. At rest, the nostrils should not be flared and just barely quiver with each breath. Normal respiratory rate is 10 to 30 breaths per minute. Babies normally have a slightly higher respiratory rate. If the animal holds its head still, equal amounts of air can be felt coming from the right and left nostrils. The nostrils should be dry to slightly moist with CLEAR mucus.

Upper Airway Problems

The signs of respiratory disease vary depending on its severity. In general, upper airway problems are less serious than those in the lower airways. The upper airways include the nasal cavity, trachea and bronchi. Upper airway inflammation may be due to irritation, allergies, infection, or abnormal growths. Irritation to the airways is commonly caused by dusty surroundings and dusty feed. The llama with a dust-induced respiratory problem will have nasal discharge and gurgling sounds from the nose. The llama still acts normal, eats well, has no fever, and may or may not cough. Obvious therapy is to decrease the dust in the llama's environment. Sprinkling down dusty areas or sprinkling down dusty hay alleviates the signs. If the hay is very dusty, change to a better hay.

Llamas, just like people, can have allergies. The offending substance may be pollen, dust, weeds, molds, feathers, and many other substances. Often the allergy is seasonal, as with a pollen allergy. Even parasites, such as lungworms, may set up an allergic response. Through testing, the offending allergen may be identified. The llamas with allergies generally act normal, eat well, do not run a fever, and tend to have high eosinophil (a specialized type of white blood cell) counts on blood tests. These llamas sound congested and usually have a clear mucus nasal discharge. The therapy is to remove the inciting cause if it can be discovered and removed. Antihistamines and changing the feed substances often help.

Nasal discharge may also be present from sinus infections, nasal cavity infections, masses, or foreign bodies. If nasal discharge is present on one side only, you may discover a foxtail, nasal bots, an abscess, or tumor on that side of the nasal cavity. These llamas tend to have noisy and difficult breathing. A veterinary examination and X-rays will confirm the diagnosis and appropriate treatment can be taken.

Infections in the upper airways also occur. Cough and/or nasal discharge are usually present. When at rest, the llama usually does not cough. However, when the respiratory rate increases, as with exercise, a cough is elicited. These llamas may be off feed or slightly depressed. A fever generally accompanies the inflammation. Your veterinarian should treat the infection with appropriate anti-microbial medicines and provide symptomatic relief for the fever, cough, and discomfort.

Lower Airway Problems

One of the most severe, but fortunately rare, diseases in llamas is pneumonia. Pneumonia is a lung infection. It can be caused by foreign materials in the lungs, microorganisms, or parasitic damage. The sick llama may or may not cough. If a cough is present, it is a painful, short, moist-sounding cough. The llama may or may not have a nasal discharge. If present, the discharge is thick white to yellow mucus. The llama is depressed with a decreased appetite. Sometimes no other signs are present except that the llama is down and depressed. If observed closely, it is distressed when breathing. Usually a fever accompanies pneumonia. Veterinary attention should be obtained immediately, as the llama's condition may deteriorate rapidly and cause death.

Tuberculosis

Although uncommon, llamas and alpacas are susceptible to Tuberculosis. Tuberculosis is caused by various species of *Mycobacterium,* a bacterial microorganism that can cause disease in many animals, including humans. Signs of disease depend on the location of the infection, with digestive and respiratory systems most commonly affected. Signs include difficult breathing, diarrhea, weakness and chronic weight loss. This disease is tentatively diagnosed by sensitivity to a tuberculin skin test. Your veterinarian must report all positive responses to state authorities. Research is underway to investigate the validity of various diagnostic tests for llamas and alpacas. The fate of animals diagnosed with Tuberculosis will be determined by governmental control programs.

Troubleshooting Respiratory Problems

In general, if you ever have a llama with a nasal discharge, difficult breathing, or a cough, watch it closely. If it is acting depressed, not eating, or in respiratory distress, call your veterinarian for assistance. If it is acting normal, is alert, and eating well, take a daily temperature. If it is elevated, call the veterinarian. If the temperature is normal, rest the animal and continue to monitor the temperature daily so that any rise will be discovered immediately. If the signs linger, obtain medical advice.

Choanal Atresia

Rarely, a llama or alpaca baby is born with a congenital abnormality of the respiratory system called choanal atresia. These babies will have respiratory difficulty from birth. They are unable to breathe if their mouth is closed. As a result, they cannot nurse and breathe at the same time. The choanae normally allow the air to pass from the nasal passageway to the throat, where it can proceed down the trachea to the lungs. With choanal atresia, an embryonic membrane or bony plate fails to rupture during fetal development This creates an obstruction between the nasal cavity and the throat. The obstruction may be on one or both sides of the nose. It may be a complete or partial obstruction. If your newborn cannot breathe with its mouth closed, choanal atresia should be suspected and your veterinarian contacted. The veterinarian will need to establish an open airway for the baby. This is done by placing a tube in the mouth or placing a catheter directly into the trachea (tracheostomy). Euthanasia is usually recommended, but surgical repair may be possible if the abnormality is not severe. However, several complications may occur including the lack of colostrum transfer, due to the baby's inability to nurse. This makes the baby very prone to any infection. Other complications are the aspiration (inhalation) of milk into the lungs, swallowing difficulties, and infection from the tracheostomy, all of which can result in pneumonia. Lastly, reoccurrence of the airway obstruction may happen post-operatively. In general, the prognosis is very poor for llamas and alpacas with choanal atresia.[1]

Ventilation

Finally, ventilation must always be considered when discussing respiratory disease. Poor ventilation is a major contributing factor to respiratory disease. When ventilation is poor, toxic fumes and dust collect in the air, as do increased numbers of harmful fungal spores, bacteria, and viruses. To keep respiratory disease at a minimum in your herd, ensure adequate ventilation in your barns.

CHAPTER 14:
DIGESTIVE PROBLEMS

Introduction

Llamas and alpacas have a unique digestive system. Their stomachs have three compartments. They regurgitate their food to re-chew and re-swallow it. This is what is commonly called chewing cud or rumination.

Rumination enables the llama or alpaca to break down its feed materials VERY efficiently. By the time all of the nutrients are absorbed from the digestive tract, the waste is formed into small pellets called feces. The pellets vary in size with the size of the animal. The color of the feces is generally green-brown when fresh, but turns blackish after drying for a while. The color, however, will vary with the type of feed ingested.

Salivation

Most diseases of the mouth or throat in llamas result in excessive salivation, foaming, or drooling. Salivation can be from nervousness or appetite stimulation, but should always be investigated to find an underlying cause when it is excessive and continuous. When investigating salivation, don't forget to consider rabies if you are in an area where rabies is present. If so, obtain professional help!

Other infectious diseases can also include salivation as a presenting sign. These include diseases which cause mouth sores such as Vesicular Stomatitis, Orf, and Foot-and-Mouth disease. The mouth sores cause irritation, which stimulates salivation. North America is free from Foot-and-Mouth disease. Foot-and-Mouth disease, a viral infection, is a serious disease in cloven-hoofed animals (such as cattle, sheep, goats, swine and deer). Affected cattle get painful blisters in the mouth and on the feet. Although the death rate is low, the disease is very contagious and spreads rapidly through the air and through ingestion. While llamas and alpacas can get Foot-and-Mouth disease, it is uncommon.

Vesicular Stomatitis, clinically resembles Foot-and-Mouth disease and occurs sporadically in North (and South) America during the warmer months. Vesicular Stomatitis is a viral disease that has been diagnosed in llamas, but primarily affects horses, cattle and swine. People who handle infected animals can also contract Vesicular Stomatitis and can exhibit flu-like symptoms which are rarely accompanied by mouth sores. Infection in animals, which is likely spread by insects or by contact with infected animals, may result in fever and blisters in the mouth, tongue, nostrils, feet and teats. The mouth sores result in salivation which is the most common clinical sign. Death rarely occurs, although the animal may refuse to eat or drink. Typically the disease is self-limiting and runs its course in about two weeks. If you suspect your animal has symptoms of Vesicular Stomatitis, immediately isolate it from non-affected animals and remember to use good personal sanitation. Then call your veterinarian who must report suspected cases of Vesicular Stomatitis to state veterinary officials. If the diagnosis is confirmed, the premises are ordinarily quarantined and travel restrictions are imposed. Officials decide when the quarantine is released, but often it is thirty days following the last sign of Vesicular Stomatitis on the premises.

Any mouth irritation may stimulate salivation. This includes cactus spines, foxtails, or any other mouth and throat injury. Teeth problems such as infected teeth, abnormal wear of teeth, and decayed teeth can be points of irritation. Occasionally new teeth erupting can lead to salivation. Certain medicines and chemicals stimulate salivation. Keep this in mind, as salivation may be the initial sign in poisonings. Many insecticides for animals, chemical insecticides for plants, and internal parasiticides contain organophosphates. In high doses these can cause toxicity and death. However, if discovered early, the malady can be reversed with appropriate drugs.

Choke is another problem that sometimes exhibits excessive salivation. Choke is the common term for an obstruction of the esophagus. It is an emergency; however, it should not be confused with choking in people where the windpipe is obstructed. Since the obstruction occurs in the esophagus – the long flexible tube connecting the throat and stomach – breathing can usually still continue. Lodged feed or some foreign object may obstruct the esophagus. The llama will appear distressed and have a sudden onset of salivation, which may be mixed with feed material.

Frequent swallowing movements are made in an attempt to dislodge the obstruction. Mild obstructions are often dislodged by the llama itself. However, these llamas should be observed closely for a couple of days for any repeated signs of obstruction. Many llamas will need veterinary assistance to dislodge the obstruction. Massaging the neck often keeps the muscles relaxed until your veterinarian arrives to help.

A few complications may occur following a choke episode. The gas in the stomach may be trapped as a result of the obstruction. In other words, the llama cannot burp. Bloating of the stomach results and needs immediate medical attention. Also, the ingested food material that is brought up with the saliva may be aspirated or inhaled into the lungs. This is a sure way to start pneumonia. If there is any chance that feed inhalation occurred, as indicated by copious salivation or a prolonged choke, the llama should be given systemic antibiotic therapy. Scarring is a complication that may occur long after the choke incident. Scarring occurs on the damaged part of the esophagus. It leads to a decrease in the diameter of the esophagus termed stricture. This stricture makes that location prone to further obstructions.

Diarrhea

Diarrhea is a condition in which the feces do not become compacted into pellets. The causes vary. Mild cases may be due to nervousness or food that is slightly rich. These cases are generally self-limiting. This is commonly seen in babies whose mothers produce rich and abundant milk. These babies act normal, have a normal appetite (nursing regularly), have no fever, no pain, and no signs of dehydration. Treatment includes a little oral Kaopectate® if the feces are liquid. The baby's bottom needs to be kept clean, as the caked on diarrhea is irritating to the skin. Diaper rash ointment may be needed if chafing occurs.

Profuse or continued diarrhea can cause dehydration and should have medical attention. The fluid that normally is absorbed into the blood from the intestine is eliminated with the feces. Signs of dehydration include decreased skin elasticity. Check for this by gently pinching the skin in the shorter fibered areas such as the chest and cheeks. Normally, the skin immediately pops back into place when the pinch is released. A dehydrated animal's skin will be slow to return to normal or remain pinched. Sunken eyes, decreased urination frequency, and a dry mouth are further signs. In a dehydrated animal, the membranes inside the mouth (gums) take longer than two seconds to return to a pink color when blanched with digital pressure. This is termed increased capillary refill time. As dehydration worsens, shock and finally death may occur. Thus, an integral part of therapy for continuous diarrhea is fluid and electrolyte administration. These fluids may be given orally, subcutaneously, or intravenously, depending on the degree of dehydration.

Diarrhea is particularly serious in young llamas. Diarrhea in a SICK baby needs immediate medical attention. In addition to the diarrhea, these babies are depressed, not nursing vigorously, and may be feverish. This may be a sign that a baby did not receive its protective colostrum as a newborn, and therefore cannot fight off infection. Since the crias lack back-up reserves, they can deteriorate rapidly. Stress can increase the speed of the illness. These babies are dehydrated and need fluids and anti-diarrhea medications.

After your veterinarian rehydrates the animal, the inciting cause of the diarrhea must also be treated. Parasites, bacteria, viruses, allergies, and excessive amounts of rich feed may all cause diarrhea. Additional tests such as blood counts and fecal exams may indicate the cause.

Many microorganisms can cause diarrhea. These infections cause mild to extremely severe inflammation of the digestive tract. For example, Enterotoxemia may manifest with diarrhea. If your llamas are not vaccinated against enterotoxemia, they may develop this serious disease. This can also occur in a baby if it does not receive colostrum, or gets colostrum from a non-vaccinated mother (or goat). Enterotoxemia is caused by bacteria. Toxins produced by the bacteria make the llamas quite ill and often cause bloody diarrhea. Unfortunately, this disease often progresses so fast that treatment is not effective. Sometimes the llama is just found dead. Vaccination is the best prevention. (See Chapter 5 for further information.)

Diarrhea may also be caused by the *Salmonella* organism. Salmonellosis is a con-tagious disease of animals and people. It is caused by bacteria. The diarrhea is started by toxins released by the bacteria in the intestine. The animal is predisposed to Salmonellosis by stress. Some stresses include transport, bad weather, indiscriminate use of antibiotics, sudden feed changes, and overcrowding. A sick animal develops diarrhea a few days after the stressful incident. The diarrhea is watery with some mucus and occasionally has blood specks in it. The animal usually has a fever and is dehydrated. If severe, Salmonellosis can result in death. Diagnosis is made by fecal culture. Use care in handling these feces, as they are contaminated by the bacteria. Fluid therapy and antibiotics are frequently needed. This disease's course stresses the importance of isolating new or sick animals to prevent the spread of disease.

Intestinal parasites such as coccidia can cause diarrhea as well as a very ill, depressed, and down llama or alpaca. Occasionally coccidia may cause neurologic signs such as seizures. It is usually subclinical (no observable signs) in the adult. In young or stressed animals, or animals without previous exposure, diarrhea is the main sign. The diarrhea may be bloody. Coccidia can be diagnosed by fecal exam. The llama will need therapy such as sulfonamide drugs to combat the organism, plus supportive fluid and electrolyte therapy. If coccidia is a problem in your herd, preventative measures should be taken. These include manure removal, pasture rotation, decreasing animal density, and preventative medicines recommended by your veterinarian. (See Chapter 5 for more information.)

Although rare, a few llamas and alpacas have been diagnosed with Johne's disease or Paratuberculosis. This is a disease of cattle, goats and sheep caused by bacteria related to those causing Tuberculosis. Llamas are not an important or vital reservoir of this disease. Infection causes chronic diarrhea and extreme weight loss. Diagnosis can be difficult. Presently, little information is known about Johne's disease in llamas and alpacas.

Exposure probably occurs by ingestion of feed or water contaminated by infected cattle, goats, or sheep. Since Johne's disease is not an extremely contagious disease, fence-line contact with domestic livestock presumably does not encourage spread of the disease. Although it may be overzealous in view of the occasional occurrence of Johne's disease, a preventative measure for llamas and alpacas is to avoid cohabitation with cattle, goats or sheep.[1]

Constipation/Impaction

Constipation refers to difficult or infrequent bowel movements. The most common cause of constipation in babies is the retention of the meconium. The initial feces, or meconium, are so hard and large, that they cannot be passed easily. The baby will strain and may roll in pain. Sometimes the only sign of a constipated baby is that it is more lethargic than it should be. An enema aids in the passing of the meconium and quickly brightens up the baby.

An impaction is a clogged intestine. The llama has a spiral colon in which the intestinal diameter greatly decreases. This narrowing is a common site of obstruction, although obstruction can occur anywhere along the digestive tract. An underlying cause for the impaction may be a parasite problem, although dried feed clumps, abnormal growths, foreign objects, and strictures may also be implicated. Have the feces checked for parasites. If the fecal is positive for parasites, deworm your llama when it is feeling better. Then, establish a frequent deworming program for your entire herd.

Signs of impactions usually include colic signs. Colic is the general term used for a llama with a bellyache. The llama is often depressed, has no appetite, may be down, and roll from the pain. Bowel movements may be scant, with hard feces, or may cease completely. The llama may spend the day straining over the manure pile. Gastric motility and cud chewing decrease.

Impactions need to be treated by your veterinarian, because the intestine may become devitalized if the impaction continues. This would result in the llama's demise. Treatment includes enemas and laxatives (mineral oil, Magmilax®) given orally. A note of caution, please have your veterinarian give the mineral oil with a stomach tube. If you give it by squirting it in the mouth, it may accidentally go into the lungs and asphyxiate the llama! General supportive care such as fluids and pain relief medications are also needed. If this basic therapy does not work, the llama may need surgery to correct the obstruction.

Prevention of all of these problems is based on a good herd health program. Deworm llamas and alpacas as deemed necessary by microscopic fecal examinations. Vaccinate the animals regularly. Keep the animal density low. Try to maintain a low stress and clean environment for your animals. Feed the animals good quality, clean feed. Make any changes in the amount or type of food very gradual. Finally, keep salt and clean fresh water available at all times.

CHAPTER 15:
THE DOWN LLAMA OR ALPACA

Introduction

A llama that lies excessively on its side or chest when sick or injured is considered a down llama. Being down is sometimes the only sign an ill llama shows. A down llama is DEFINITELY a cause for concern. First of all, for the novice owner, it should be noted that llamas normally rest and chew cud lying down. They are usually sternal (on their chest), but frequently lie flat out on their sides. This is normal behavior. As you get to know your llamas' behavior, you will notice that they have favorite places to lie down where they sunbathe, nap, and chew cud. If they are chewing cud, it is obvious that they are normal. Thus, a normal resting llama should not be confused with an ill llama who spends the majority of the day down. If you are in doubt as to your llama's down behavior, stimulate it to rise. If it rises normally and proceeds to eat, void, and act like a normal llama, it was probably just resting.

Down behavior should also be viewed in light of the particular circumstances. When some llamas are asked to do a task that is strenuous, distasteful, or intimidating, they show protest by lying down. Often a scared llama will go down in submission, which occurs frequently in chutes. With practice, you can learn how to get a protesting or submissive llama to rise. Running towards it quickly to startle it or pulling its hind legs out back behind the llama will encourage it to rise.

If you have a down llama loose in its pasture that rises normally upon stimulation, but then wants to go down again immediately, or totally refuses to get up, you have a problem. Being down or staying down is a sign of depression and/or pain in llamas. Keep this in mind if your llama is near her due date, as uterine contractions are painful and you may have an ensuing birth. However, if this behavior continues throughout the day, and no baby arrives, it could be a sign of dystocia. Uterine torsion, or twisting, also causes pain in females and while uncommon, typically occurs in late pregnancy. Obtain veterinary assistance. (See Chapter 16.)

Colic

A very common reason for a llama to be down, when it is capable of rising normally, is a bellyache or colic. These llamas don't want to eat, and may or may not drink water. They just prefer to lie around most of the day. In addition to their reluctance to rise, the llamas may roll and frequently change positions showing general discomfort. They may grind their teeth and may stent (stiffen) their belly when you touch it. The causes for belly pain are numerous. One common cause is impaction. (See Chapter 14.) These sick llamas have hard and dry feces that are scant and abnormally shaped. They need laxatives and maybe an enema, as determined by your veterinarian. Once the impaction is passed, these llamas should soon be up, eating, and back to normal.

Diarrhea may also be painful to the llama and cause it to be down. The muscle walls of the intestines may be cramping or contracting quickly, causing pain. Also, if the diarrhea is caused by an infection, the llama may be down due to generalized depression and illness. (See Chapter 14.)

On rare occasions, a male llama or alpaca may show signs of colic, not due to stomach and intestinal pain, but rather urinary pain. The urethra, which transports urine from the urinary bladder through the penis when urinating, can become partially or completely blocked with cellular debris and mineral crystals. Besides exhibiting pain, these males typically strain to urinate. Relief of the obstruction is essential to the health of the entire urinary system and hence the animal. Contact your veterinarian for treatment.

Ulcers

Another fairly common cause of belly pain is ulcers. An ulcer is an erosion of the lining of the stomach or duodenum (intestine) wall. Llamas under stress seem to be prone to ulcers. Ulcers seem to occur more often in llamas fed high grain, low fiber diets. Some drugs may irritate the stomach directly, causing ulcers, while others initiate excessive chemical secretions which may result in ulcer formation. These animals do not want to eat and are down most of the day due to the discomfort of the ulcer. They may be anemic from significant blood loss as the ulcer bleeds. Generally, few feces are passed.[1] If present, the diarrhea may be black and tarlike due to the digested blood. If the ulcer perforates the wall of the stomach, severe contamination of the abdominal cavity occurs and this generally results in death. Due to the potential consequences, if an ulcer is suspected, the llama should receive prompt veterinary treatment. If an inciting cause can be found, such as a stress-causing environment or the usage of harsh oral medicines, it should be corrected. Your veterinarian may give Cimetidine (Tagamet®) injections or other acid-reducing drugs such as omiprazole. Cimetidine is an anti-acid producing drug. Blood transfusions may be needed if the animal is severely anemic. Always consider the possibility of an ulcer for a llama who is down, in pain, and not eating, when other causes cannot be found.

Infection and Organ Failure

Any infection, from an infected cut to pneumonia, may cause a llama to be sick enough to stay down for most of the day. Any organ failure (kidney, liver, heart) weakens the llama and makes it depressed, causing it to decrease or stop eating and to be down. Fatty Liver Syndrome, or Hepatic Lipidosis, can make an animal quite sick. While not commonly recognized in all regions of the United States, it is likely the most common form of liver disease seen in llamas and alpacas. While the progression of the disease pathway is not clearly understood, some factors are believed to predispose these animals to Hepatic Lipidosis. These factors include a change in diet, being underweight or overweight, high energy demands such as late gestation or milk production, a change in the environment, and any type of stress.[2] This condition results in the accumulation of fat in the liver which impairs the liver's normal functions and can have lethal consequences.

Infections, metabolic disorders, or organ failures can be diagnosed by veterinary examination, which may include blood screening tests, urine samples, and fluid samples. If your down llama has recently given birth, there may be complications such as mastitis, uterine infection, uterine tears, toxemia, blood clots, or mineral imbalances. The female needs immediate medical attention!

West Nile Virus Infection

West Nile virus was first detected in the United States in 1999. The virus is transmitted to animals and humans through the bite of an infected mosquito. Mosquitoes get the virus by feeding on infected birds, which appear to be the reservoir for the virus. Infected animals, including llamas and alpacas, may or may not become ill. Those that become ill have fever and show neurological signs such as incoordination, head tremors, convulsions, difficulty or inability to stand, and death. Treatment of an ill animal is only with supportive measures which may not be enough to save it.

It would be beneficial to animals and humans alike, to control mosquito numbers on your ranch. During mosquito season, steps must be taken to eliminate any standing water. This is where mosquitoes breed and the larvae develop. Drain or discard any items (including old tires) that can collect water. Make sure roof gutters drain properly. Maintain permanent ponds with mosquito-catching fish.[3] The necessity of vaccination in llamas or alpacas, horse vaccine has been used in camelids. If you and your veterinarian choose to use a horse vaccine to establish immunity, it makes sense that the initial vaccine and the booster are administered prior to mosquito season.

Injury/Paralysis

Another possibility for a down llama is one that wants to get up, but cannot do so. Often this means a severe injury such as a fracture. First, check the animal for any signs of injury such as blood, wobbly leg, swelling or bruising. This may make the diagnosis obvious. The injured animal should not be forced to move and you should call your local veterinarian and describe the injury to obtain further advice. Clean any injury gently and keep it protected from dirt and debris until assistance is available. If you are out on a trail, in the middle of nowhere, and suspect a fracture, or have a severe injury, keep the animal as comfortable as possible. This means removing the pack, offer him water, and support the injury if it is a leg, so that it will not worsen. A splint with plenty of padding will suffice. (See Chapter 11 regarding bandaging fractures.) Move the animal slowly, at its speed, not yours, to the trailer. A llama can walk on three legs.

What if your llama is down and cannot move at all? It cannot even bend its legs. **This is an emergency. Call for medical assistance immediately.** Your llama may be so severely

ill that it cannot move. It may have a neurologic injury or infection such that the messages cannot get from the brain to the muscle to make them move. The llama may be partially or completely paralyzed. Paralysis may also be caused by a tick bite. In an area where ticks prevail, immediately start checking for and removing ticks. Ticks seem to like the tender skin of the mammary glands, prepuce (penis sheath), belly, chest, face, in the armpits, and under the tail. The animal should also be treated with an insecticide and/or parasiticide for any ticks you missed. This, plus supportive treatment should be done under doctor's recommendations.

The meningeal worm (*Parelaphos-trongylus*) should be considered as a possible diagnosis for a down llama in the eastern United States and Canada where white-tailed deer are found. The llama accidentally ingests infective *Paraelaphostrongylus* larvae in a snail. The worm then migrates up the llama's nerves to the spinal cord and ascends towards the brain. Serious inflammation occurs along the spinal cord in response to the worm migration. Signs begin as rear legs weakness and uncoordination and can lead to complete paralysis and death. Your veterinarian should be called as soon as any signs of uncoordination are seen so that prompt treatment can begin to try to save the animal's life. Prevention includes fencing off llamas from deer and decreasing snail populations around llamas.[4] (See Chapter 5.)

There are instances where the llama cannot or will not rise, and veterinary services may be miles away. Youngsters may be carried to the trailer, but older llamas provide quite the challenge. First, get the trailer as close to the animal as possible. Then, get large towels or blankets to use as a sling. Place the sling under the animal's chest, just behind the front legs. If necessary, you'll have to dig a little under the chest to get the sling under the llama. Place another one just in front of the hind legs. Then, a person should be on each side of the sling with the strongest people and main force being on the front sling to lift the animal. If the llama has no head or neck control, make sure another person supports the head and neck too. The llama can be carried to the trailer. Alternatively, you can slide the llama onto a large piece of plywood or tarp, pull it or carry it towards the trailer, and slide both the plywood and the llama into the trailer. Once in the trailer, it can be supported with hay or straw bales on either side. Now you can take your llama to assistance.

You must get to know your llama's normal habits to recognize the abnormal. If a llama is lying around more than usual, don't hesitate to investigate the problem, especially when the llama is alone and all of its buddies are elsewhere in the pasture. Most of the time, there is a problem when the llama is down for most of the day. Since llamas are so stoic, the fact that they do not want to rise means something is very wrong with them. Discovering the problem as early as possible will increase the chance of a successful recovery.

Troubleshooting Down Llamas and Alpacas

1. **Down, but gets up and stays up when stimulated. Acts normally.**

2. **Pregnant llama near her due date who spends excessive amount of time down (more than she has in the last few weeks).**

3. **Down llama; alert and wants to rise but cannot.**

4. **Down llama; gets up when stimulated, but sits back down immediately. May have a fever.**

5. **Down llama; gets up when stimulated, but sits back down immediately. No fever.**

6. **Down llama; cannot move at all.**

1. Probably resting. Observe for any other abnormal signs.

2. May be in labor. Observe for birthing. If behavior is continuous and no baby arrives, call your veterinarian.

3.
   A. Observe for injury. May have a fracture. Do not move llama unless necessary. Treat any obvious wounds. Seek veterinary care.
   B. If in a tick area, check for and remove ticks. Keep the animal comfortable. Call veterinarian for supportive therapy.
   C. If in a white-tailed deer area, consider the meningeal worm; especially if signs were preceded by rear leg uncoordination. Seek veterinary care A.S.A.P.

4.
   A. Hyperthermia possible. Provide heat relief. See Chapter 12.
   B. Infection, toxemia, or organ damage may be present. Seek immediate veterinary attention.

5.
   A. Always call your veterinarian; usually an emergency especially in youngsters). May be a bellyache if scant, dry feces present. May be an impaction or constipation. Try an enema.
   B. Diarrhea present. Isolate, offer water, and call veterinarian to determine cause and treat appropriately.
   C. Organ failure or infection. Veterinary exam is needed for diagnosis and appropriate therapy.
   D. Evidence of injury. Treat wound. Give pain killers as needed under orders from your veterinarian.

6.
   A. Check for and remove ticks. Call veterinarian.
   B. If hyperthermia is suspected, provide heat relief. If hypothermia is suspected, provide cold relief. Then, obtain veterinary assistance.
   C. The llama has feeling present in the legs. This may be a severe injury, infection, mineral imbalance, or organ failure. Seek immediate veterinary care. If down animal is lying on its left side, consult with veterinarian whether it should be moved onto its right side to facilitate expulsion of stomach gas.
   D. No feeling is present in 1-4 legs. May have neurologic injury (broken back), or infection. Do not move llama until you consult with your veterinarian.
   E. Consider selenium and vitamin E deficiency if you are in a selenium-deficient region.

CHAPTER 16:
REPRODUCTION

The Male Llama and Alpaca

Normal male reproductive anatomy includes two testicles held in a scrotal sac on the animal's posterior. Although a male with only one external testicle may be able to impregnate a female, he should not be used as a breeding stud, as this testicular trait is likely to be inherited. It should be noted that the llama can move its testicles closer or further away from the body to maintain optimal temperature for the sperm. Don't panic if your male appears to have no testicles on a cold day or during a moment of extreme apprehension!

The prepuce is the sac holding the penis. It normally points backwards and can be observed when the male urinates. When sexually aroused, the prepuce points forward, and

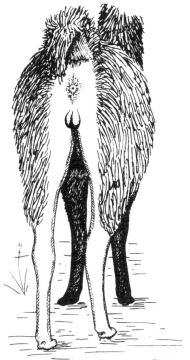

Figure 16.1 **NORMAL MALE EXTERNAL ANATOMY**

the penis extends from the prepuce. This penile extension cannot occur in a young llama or alpaca because of a normal connection of tissue between the penis and prepuce. This point of attachment disappears as the animal matures. The male llama or alpaca reaches full sexual maturity at about 2-1/2 to 3 years of age.[1] At this time he actively produces sperm as his sexual drive increases. However, due to the chance of an early "bloomer", it is wise to separate the young males from the females by one year of age or better yet, at the time of weaning. When he reaches puberty and breeds an interested female, he actively pursues her, attempting to mount her until she sits down in the kush position. Then, he sits on top of her and maneuvers his penis into her vulva. Breeding lasts 5 to 45 minutes, with average being 20 minutes.

serted and through which the baby passes. Internally, the vagina leads to the cervix which is the opening into the uterus. The llama's uterus consists of a body and two uterine horns, the right and left. The baby develops in the uterus with most llama pregnancies occurring in the left horn. The oviducts or fallopian tubes carry the eggs from the ovaries to the uterus and are the site of fertilization. The female reaches sexual maturity at one to three years of age, although some cases of pregnancy occur in females just under one year of age. It varies with females, but most can conceive by 18 months of age. In general, she should be 60 to 70% of her adult estimated weight before breeding to allow her to mature fully before channeling her nutrients to-

Figure 16.2 NORMAL BREEDING POSITION WITH MALE ON TOP OF FEMALE.

The Female Llama and Alpaca

Externally, the female genitalia consists of a vulva, which is about 1 to 1-1/2 inches long. (Figure 16.3) The vulva contains the opening into the vagina through which the penis is in-

wards reproduction, and to decrease the chance of birthing complications due to the small size of her immature pelvis. Most females are mature enough to safely breed at 18 months of age. A large and early maturing female could be bred earlier but she should be at least one year of age and over 190 pounds to provide a reasonable measure of safety. Similarly, most alpacas are first bred at 14 to 18 months of age.

Choosing a Mate

If you desire your female to reproduce and she is sexually mature, an appropriate mate must be chosen. The male should be selected with the overall goals of your breeding program in mind. Your program may dictate that the offspring be a good packer, or a pet or have abundant wool or be a show animal. Based on your priorities, male prospects can be investigated. Several factors are involved in choosing the most suitable male. If the male is not owned by you and you must pay a stud fee, the cost must be within your financial means. Depending on your situation, aspects such as the male's reputation, show and performance record may be evaluated. All prospective mates should have good conformation, move freely (unless he has suffered an injury), be even-tempered, free from genetic abnormalities, and excel in qualities where your female does not. If he has previously bred females, his reproductive abilities should be investigated; i.e., did he impregnate the females and were live normal offspring produced?

To choose the best possible match, you must first objectively evaluate your female in terms of her good and poor qualities. For example, your particular female may have an excellent disposition, good quality fiber, large brown eyes, sturdy back, and yet her front legs are knock-kneed and she was born with slightly short ears. You must decide if her strong points outweigh her weak points for the goal of your llama business or hobby. If you feel that this female meets your goals and standards and that her attributes are important to your breeding program, you must select a male who has complimentary good qualities and very strong qualities where hers are poor. In this example the male should also have a good temperament, fiber and back, but MUST have longer ears and straighter legs, and hopefully have proven that he passes these traits on to his offspring. Several of his traits

must be evaluated. just as you do not choose your females based on one trait alone, you also must not choose the male on only one trait. For example, the male should not be chosen ONLY because of his pedigree or ONLY because of a show championship or ONLY because he has good quality fiber.

Upon evaluation you may realize that a particular female does not possess the traits you desire in your future herd. In this case you should consider selling her and buying a new and more appropriate llama. This approach will gradually increase the overall quality of your females and thus lead to a higher quality offspring. If you find upon evaluation that your female's poor qualities outnumber her strong points you should not breed this llama at all. If she has that many weaknesses or problems, at least some of them are likely to be inherited. Also, usually one genetic abnormality is accompanied by other genetic abnormalities. These qualities should not be passed on through generations. The inferior offspring would harm your own reputation and the integrity of the entire llama population.

It is the responsibility of the llama and alpaca industries to breed for healthy and high quality animals and to prevent the perpetuation of harmful traits that may be inherited. This is indeed a challenge to owners as significantly fewer variations of inherited traits exist in the llamas in the United States. This is due to the limited populations of the "founding" llamas, leading to a fairly small gene pool in the current population of llamas and even smaller alpaca gene pool. This decreases the chances of diversifying genetic combinations when choosing a mate for your animal. Animals with diversified genes tend to be stronger, more vigorous and healthier. Breeding of distantly related llamas and alpacas may be unavoidable with current populations in North America. However, breeding of closely related animals greatly increases the chance of expression of a genetic abnormality.

Inbreeding, or breeding of two very closely related llamas such as a mother and son, decreases the variety of genes in their offspring. Thus a harmful or even deadly gene

101

will express itself because no normal gene is present to "cover" it up. Some examples of these genetic abnormalities that are expressed with close breeding are choanal atresia (see Chapter 13), lack of a rectal opening (Atresia ani), fused toes, and stubby ears (see Appendix V). Also, inbred individuals are often weaker, reproductively inferior, and may have other anatomical and physiological abnormalities. With all of its potential harm, inbreeding should be avoided by the novice llama owner. However, occasionally an inbred individual does not have the above abnormalities and has many similar GOOD quality genes. Due to the homozygosity of the genes (homogeneous set of genes), this llama's traits have an excellent chance of being passed on to its offspring. These are the superior genetic animals that give their "look" to their offspring.

If this superior animal's genes are desired, linebreeding may be practiced to perpetuate this animal's good traits. Linebreeding is the mating of animals that are related to this one common superior ancestor. However, this is still breeding of related animals and should be done with extreme caution, because the same harmful outcomes can still occur as with inbreeding. Many knowledgeable llama authorities contend that neither inbreeding nor linebreeding should be practiced in the United States because these practices limit the genetic diversity most breeders are trying to develop.

The ultimate proof of a male's or female's genetic quality is expressed in their offspring. The physical appearance of a llama is the result of its inherited traits (genes) and environmental factors such as nutrition, injuries, amount of exercise, parasite load, and overall health. The best way to determine which of those traits are genetic is to see what he or she passes on to the offspring. For example, if the llama has stubby ears from an unknown cause and produces offspring with stubby ears, it is probably a genetic trait.

For logistical reasons, a male can demonstrate his genetic abilities easier than a female can. A male can have several offspring in a year's time, whereas a female can only have

one. If a male has available offspring, they should be examined closely to see if they also have the good qualities that the male possesses. Any one offspring is the result of the genes received from its mother and father. Therefore, if it is possible to see the offspring's mother to evaluate her good and inferior qualities and if she has poor qualities, you can hypothesize as to whether the male's genes may have improved upon these traits in their offspring. If an offspring has a genetic abnormality, it is possible that the trait came from either or both parents. Since the responsible parent is often unknown, this increases the risk factor involved in choosing this particular male as a mate.

Genetic traits should be analyzed by studying the traits of the llama's ancestors. A thorough study of a prospective male's pedigree will reveal how closely related he is to your female, the quality and reputation of his ancestors, and whether the particular traits passed on from those ancestors are desirable traits in your breeding program.

Finally, if the male is not your own, terms of a breeding agreement must be reached. Although verbal agreements are common, a written contract is better for protection of both involved parties. Besides financial terms, details regarding the potential of infertility in the male or female should be worked out. For example, if the female is infertile, might you breed another female to the male or do you get a partial or full refund? Often a booking fee is a non-refundable portion of the stud fee and the remainder is paid upon birth of the baby. Terms of the female's health care management and board also must be discussed. The owner of the male may request an examination stating that the female is free from any communicable diseases. Even unpleasant details such as the terms involved if the male dies or becomes unable to breed, or the female dies before the baby is born should be addressed. Finally, any warranties must be agreed upon. Is pregnancy guaranteed or is a live baby guaranteed? If a live baby is guaranteed, determine at what point the warranty expires. Often a live baby guarantee means that the baby is born alive and stands and nurses. After this point, the fate of the baby is usually not the responsibility of the party owning the male. If the baby is not born alive or dies before the warranty expires, terms must be agreed upon as to whether a rebreeding is free, if another female may be substituted, or if the money is refunded. Although most breedings and births are uneventful, it is important to have all of these details agreed upon BEFORE you breed your female.

Breeding Methods

Pasture Breeding:

Many people who have only a few females and one stud like to keep the male in with the females all the time. This has the advantage of making fairly sure that whenever a female is open, she will be bred. Sometimes however, familiarity breeds disinterest, and in this case the male's interest may return if he is separated from the females for a few days to a week or more. Pasture breeding also makes management of the herd easy, since separate quarters are not needed, and fences do not have to be exceptionally strong to keep the male away from the females. On the negative side, it means that unless a breeding is actually witnessed, and a careful watch kept ALL the time, you may have no idea when a breeding occurred, and thus when delivery could be expected. Younger females may also be bred before they are really old enough or to an inappropriate male, without anyone knowing. In addition, if there is a reproductive problem, diagnosis will be harder due to the lack of dated specific history.

In South America, where high reproductive rates in llama herds are more desirable than known parentage, an alternative method of pasture breeding is used. Two teams of two males are pastured with the herd of females, with each team alternately spending one week with the females, and one week resting. The short period of time with the females, the presence of a competing male, and the rest and separation from the females combine to keep breeding interest high. In this situation, it is very unlikely that a female will remain open long. On the negative side, the father of a baby sired in this way could be any one of four males, and the males themselves may be more likely to fight and be injured as they compete for breeding rights.[2]

Hand Breeding:

Hand breeding is only workable when males and females live separately and it is much more demanding logistically. There must be at least two independent sets of living arrangements, depending on the number of males and females involved, and wear and tear on the fences may be greater. However, it has a number of benefits to offset the disadvantages.

Hand breeding involves haltering both the male and female, and leading them to each other at a time and place chosen by you. It requires two people, who must be strong enough to stay with the llamas and separate them if it proves necessary. This might be the case when a female shows she has ovulated or is pregnant by rejecting the male, while the male is still very interested. In this case, the people involved might be well advised to wear raincoats, since the female may spit quite a lot in her desire to tell the male off!

Hand breeding allows the owner to keep a very specific reproductive history, so that delivery dates can be calculated, and if a reproductive problem does exist or develop, it can be easily diagnosed. It also makes it possible to schedule birthing to avoid births during midwinter or midsummer months, or as appropriate for your specific area. Males who are very aggressive breeders can be restrained somewhat during hand breeding. Females being bred for the first time are occasionally unaware of what it all means and what they are supposed to do. If you are doing hand breeding you may be able to prompt them a bit. Older or more frail females can be protected from the run around, too. Hand breeding also allows the flexibility of keeping a group of females living together, but with each female bred selectively to a different male. On the negative side, it is more demanding logistically, since it requires two people. It also means that breeding checks must be made regularly, since a female who did not get or stay pregnant will not have another chance until you intervene.

Corral Breeding:

This is a compromise between pasture and hand breeding. The male must be living separately, and females can be introduced to him for brief periods to check the response. If it turns out to be negative, and the female is not interested, she will either have to tell the male off herself, or be released. Some bred females, when faced by an amorous male who won't take NO for an answer, will try to escape by jumping a fence and this is definitely to be avoided. Occasionally, if several females are put in with a more well-behaved male, the male will be put off by the first one or two adamant refusals, and will go off to a corner to sulk, showing little interest in other new arrivals. While corral breeding does not provide the same degree of control as hand breeding, it does allow one to be sure whether a breeding took place and know exactly how the two llamas responded to each other.

"Escape" Breeding:

While it is to be avoided if at all possible, there will be occasions when a gate didn't get latched, or was broken, or a male jumped a fence or partition. You arrive at the llamas' pasture to find that there is an extra llama in with the females, and you don't know who may have been bred. Once the male or males can be caught or herded back where they belong, take the time to inspect each female. Look for matted hair over the rump and on the sides, where the male's legs would have rested. Check the tail and rump for wetness, matting or even a small amount of fresh or dried blood. These may be indications that the female has been bred. Check the reproductive records of the llamas in question. Often, if the male actually had to jump or crawl under a fence to get to a certain female, it is a strong indication that she was open, and this may be indicated in the records. Of course, if a number of hours or days have passed before you realize what happened, your chances of guessing right are low. Still, enter the escape in your records, and if a female delivers

a baby 345 days later, the probability is high that this was the effective breeding. Sometimes a male will breed two or even three females during an escape, and it is possible that more than one baby may result. If there is a possibility that a father-daughter or mother-son breeding occurred, it is probably a good idea to discuss abortion options with your veterinarian, since offspring from such a union are not likely to be very desirable. Another possible hazard of "escape" is that, if a female is near the time of delivery (within a month or so), her hormone levels may be changing. This may make her especially interesting to males. She, however, is not interested in breeding, and if the male is persistent, there is some risk to the mother and fetus, due to the male's efforts to get the female down.

Normal Breeding

Behaviorally, female llamas show sexual interest by assuming the kush position when aroused by the sight, smell, or sound of a male llama. When a non-pregnant female is presented to the male for breeding, she normally sits down in a kush position within a few minutes. The male then positions himself on top of her. Most females remain in a kush position throughout the breeding, but some may roll onto their sides and could cause an inexperienced male's penis to become dislodged. If the female has a woolly tail, it may be a good idea to wrap it, trim it, or put a sock over it to keep the hairs out of the way before taking her to the male.

Female llamas and alpacas only ovulate (release an egg) upon sexual stimulation or by the use of certain drugs. The penis intromission during breeding normally stimulates the ovulation of a female having a fully mature egg in her ovary. Sometimes it takes more than one breeding over several successive days to stimulate the ovulation. Rebreeding 12 to 24 hours after the original breeding helps to insure ovulation.[3] Once the female has ovulated, she will normally refuse the male by showing dislike, running away, spitting, and refusing to sit down when exposed to him. If there is some question as to whether your female has

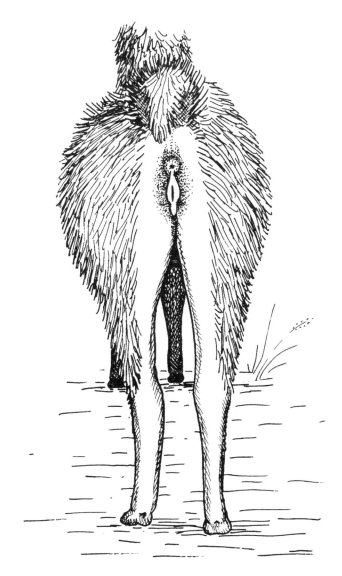

Figure 16.3 *NORMAL FEMALE EXTERNAL ANATOMY.*

ovulated, a blood test for the level of progesterone can be run one week postbreeding refusal. Progesterone is the hormone which supports pregnancy. Following ovulation, the remaining follicle cells in the ovary (corpus luteum) produce progesterone. Thus, a high blood progesterone level (greater than 1 ng/ml from the laboratory) indicates ovulation. If she ovulates but does not become pregnant, the follicle cells (corpus luteum) will stop making progesterone by about 14 days postbreeding and she will likely be receptive by 21 days. She will no longer act pregnant and will accept the male for breeding.

If the female is pregnant, she will continue to make the progesterone past 21 days and throughout the pregnancy. Therefore, behaviorally, she will usually not accept a male, and if a blood progesterone is run, it is generally

above 2 ng/ml. Levels of 1 to 1-1/2 ng/ml are questionable and should be rechecked. Females with levels below 1 ng/ml are generally not pregnant but these low levels have been observed during normal pregnancy due to progesterone variations over time.[4] The female may also be checked for pregnancy by an ultrasound exam at about 20 to 28 days of pregnancy or thereafter. Alternatively, a rectal palpation can be done by your veterinarian after about 40 days of pregnancy. Rectal palpation and ultrasound, which is done per rectum, must be accomplished with adequate restraint and a cooperative female to ensure safety and minimize the risk of any rectal damage. Non-invasive ultrasound examination through the left abdomen can also be accomplished beginning at day 50 post-refusal, unless the llama is too fat. After day 90, this is

105

best accomplished on the right side, until term.

For unknown reasons, embryos may die around 45 days of pregnancy. These may be caused by abnormal or perhaps right horn pregnancies. Therefore, it is extremely important to reconfirm pregnancy after 45 days by one of the above methods. External signs of pregnancy are difficult to detect and should not be relied on.

Rebreeding

Some llamas and alpacas become pregnant very easily and carry full term pregnancies year after year. Healthy mothers can have babies yearly. Ten to twenty-one days following parturition, the female seems to have good fertility and could then be bred. Some females should be given a few additional days for body repair and replenishment if the pregnancy and delivery were a noticeable burden on her body. Rebreeding should be postponed at least three weeks following a dystocia, because the uterus needs additional time for repair. The length of postponement varies with the severity of uterine damage and should be discussed with your veterinarian.

Infertility

The llama or alpaca who cannot get or stay pregnant is a major concern of animal breeders. The causes of infertility vary. The animal may have an anatomical or physiological abnormality and be unable to get pregnant, or she may have a mild treatable problem. The following is a troubleshooting list based on signs the "infertile" female is showing.

1. **The female llama continually sits down for the male for breeding every day or every other day. She is probably not ovulating.**

   A. Is she a very young female? If so, perhaps she is not yet sexually mature. Allow her to continue growing and maturing for two to four months and try again.

   B. Is she a large female being bred to a small male? He may not be able to adequately stimulate her to ovulate.

Use a different male, or hormone therapy as below.

   C. She may not have had a "ripe" follicle at the time she was bred. Repeat the breeding cycle in one to two days.

   D. Follicle cysts occur in llamas and will cause a failure of ovulation. These can be treated with hormone therapy. GnRH and HCG both work well to induce ovulation in llamas. Both are injectable hormones and should be given under the supervision of your veterinarian.

   E. Poor general body condition can affect fertility. Causes include nutritional inadequacies, disease, parasitism, obesity, and extremely heavy milk production. Remedy the underlying problem.

   F. If none of the above work, consult with your veterinarian. The llama may have some congenital abnormality, especially if she is three years old or more and has never had any babies.

2. **The female is ovulating but does not get pregnant. This female sits down and is bred by the male. She says "no" to him for 14 to 21 days, but then accepts him.**

   A. If no discharge from the vulva or other external signs of illness are present, try again at 21 days. If you try rebreeding for three consecutive series and no pregnancy ensues, consider further evaluation of both the female and the male llamas. Eliminate the possibility of the male being the source of infertility. If he is not impregnating any females, consider having a semen evaluation done by your veterinarian. If desired, you might also try using a different male. A male who is usually fertile may have very low sperm counts during hot weather, as sperm are very sensitive to heat.

   If still unsuccessful, you should have a thorough veterinary reproductive evaluation of that female. As this should include an examination of the internal and external reproductive structures, sedation may be necessary. A rectal and vaginal examination, and an ultrasound should be done, if possible, to feel and see the internal reproductive organs. Further tests may also be indicated depending on the findings such as a uterine biopsy to check the condition of the uterus and/or a uterine culture to check for the presence of an infection.

If the female has NEVER had any previous babies, has no vaginal discharge, and is still not pregnant after several attempts, she may have a congenital abnormality that may be discovered during the reproductive examination. She may have problems such as an abnormal cervix, only one uterine horn, an imperforate hymen, or small nonfunctional ovaries causing abnormal, or no follicle, development, and/or insufficient amounts of important ovarian hormones. In fact, she may be so abnormal that she is not completely female. Hormones such as FSH and GnRH may be tried to treat those llamas with small ovaries, but generally have a very low success rate. An imperforate hymen may be repaired surgically. The ethics involved in repairing or treating these females should be considered first as these traits may be inherited. (See Appendix V.) In general, most of the other congenital abnormalities result in a sterile female.

B. If the female HAS had previous babies, and three series of breedings result in no pregnancy, she may have a uterine infection. Uterine infections are fairly common in llamas. The llama often obtains the infection from contamination at the previous birthing or during the present breeding efforts. Vaginal discharge is usually not evident. This female should

have a full veterinary exam as above. If indeed an infection is found, it should be treated aggressively. Treatment may include systemic antibiotics and uterine infusions, depending on the type of infection.

During the reproductive exam, other abnormalities such as cysts, abscesses, or tumors may be discovered. Also, the llama may have some scarring or adhesions along the reproductive tract preventing the sperm from arriving at the egg. These may be found upon exam or biopsy. In addition some older females may have hormonal imbalances that impair fertility. Depending on the problem, she may or may not be able to have babies in the future.

C. If the female has discharge present from the vulva, and is not pregnant at the FIRST 21-day check, do not try to rebreed as above. Have your veterinarian examine her first. Discharge is usually indicative of infection. The longer an infection goes, the harder it is to clear it up with satisfactory results.

D. If the llama is in poor body condition, remedy the problem before proceeding with breeding. (See 1.E.)

3. **The female llama gets pregnant, but cannot hold it past 45 to 60 days. This is called early embryonic death.**

A. This may be "normal" attrition or die-off. The embryo may not be implanting correctly in the uterus (right horn pregnancy) or may have some abnormality itself. Rebreed and try again. After a couple of unsuccessful efforts, you might try a different male.

B. The female may not have had enough progesterone production from the follicle cells or corpus luteum to maintain pregnancy. Rebreed, try again, and measure several progesterone blood levels. If all are low, but she is behaviorally pregnant, or positive to rectal and/or ultrasound ex-

107

amination, consider progesterone supplementation. Consult with your veterinarian as to the amount and form of progesterone that would best suit your needs.

    C. If the animal is in poor body condition, remedy the problem before proceeding with breeding.

    D. If you have tried the above methods unsuccessfully, or if she has any vaginal discharge, a thorough veterinary reproductive examination is in order. (See 2.A.)

4. The female llama has not been previously bred, and will not accept a male at all.

    A. The most common cause of a female llama not accepting a male is that she is pregnant. Even if you think that she has not been exposed to a male, have her checked for possible pregnancy. You may be mistaken.

    B. She may not like that particular male. Some females are choosy and will accept another male more readily.

    C. She may have a "false" pregnancy where she is producing progesterone. She is not pregnant, but she thinks she is. This could occur as a result of "across the fence flirting" causing spontaneous ovulation, or from actual breeding. If it is confirmed by examination that she is not pregnant, the corpus luteum making the progesterone can be lysed or killed with prostaglandin drugs. Consult with your veterinarian and use caution because some prostaglandin drugs may be dangerous to use in llamas and alpacas.[5]

    D. She may have an infection or some abnormality. Have a complete reproductive work-up as in 2.A.

    E. If the llama is in poor body condition, remedy the underlying problem before proceeding with breeding.

5. The llama gets pregnant, but aborts the fetus.

    A. Causes of abortion are numerous. Poor body condition in the mother, excessive stress such as extreme environmental temperatures and physical trauma may result in abortion. Ponderosa Pine needle consumption may cause abortion in other species and may account for some abortions in llamas. Also, any fetal abnormalities often result in abortion. Twins, which are rare in llamas and alpacas, frequently abort due to overcrowding in the uterus. Repeated abortions of unexplainable causes may indicate that the llama is not reproductively fit and should be used for some other purpose.

Unfortunately, the diagnosis cannot always be made when an abortion occurs. However, the fetus should be saved for examination and diagnosis should be pursued in case an infective or toxic cause is found. Sometimes an abnormality or infection is found that could affect the aborting mother's well-being, her future pregnancies, or other llamas' health. Infective causes of abortion include Leptospirosis (See Chapter 5), Brucellosis, Toxoplasmosis, and Chlamydiosis.

    B. Brucellosis (Bang's) is caused by a bacterial infection. People can be infected by the *Brucella* organism causing an illness known as Undulant Fever. In animals, it can cause an infection resulting in abortion. This disease has not been diagnosed in llamas in North America, but has in South America. Most of these cases seem to be associated with exposure to other carrier or diseased species, such as cattle or goats.

    C. Toxoplasmosis is a widespread disease which occurs in people and all domestic animals. It is caused by a single-celled organism called *Toxoplasma gondii*. Much publicity regarding the disease concerns a harmful infection of cat owners who are pregnant. A mother acquiring toxoplasmosis encounters the possibility of causing harm to her unborn human baby. Transmission of the organism in people and animals includes congenital infection, raw or undercooked meat, or cat feces. The cat can acquire the infection from eating infected meat or from infected feces of other cats. Toxoplasmosis rarely causes clinical disease in cats. When it does, it is usually seen in kittens. The infected cat then passes the organism in its feces. The cat may defecate near llamas' feed and water sources, thereby offering a source of infection to the llamas. Toxoplasmosis could be a cause of abortion where cats are present. Until more research is done on llama toxoplasmosis, cat populations should be kept to a minimum around the breeding llamas. If cats are a MUST, feed the cats only commercial cat food or thoroughly cooked foods. Prevent the cats from roaming where they can eat wild animals. Also, have your cat's feces checked periodically for the organism (although rarely positive except in kittens) and maintain a mature cat population.

D. Chlamydiosis: *Chlamydia* microorganisms have been implicated as abortion-causing agents in llamas and other species. Chlamydial infections tend to cause abortions late in pregnancy or may cause premature, weak, or stillborn babies. Your veterinarian may submit laboratory samples from the placenta or aborted fetal tissues and/or paired blood (serum) samples from the mother to aid in diagnosis. Ensuing therapy and sanitation should be discussed with your veterinarian.

Parturition

The average llama is pregnant for 340 to 345 days. Similarly, the alpaca is pregnant for 335 to 360 days.[6] However, pregnancy can range from 330 to 375 days, with some llamas normally going a full year with each pregnancy. Parturition is the medical term used for birthing. Some llamas do show external signs of pregnancy as the parturition time nears, and some do not. Most llamas' mammary glands will enlarge one to six weeks before parturition. However, it is not unusual for mammary enlargement not to occur until after the birthing, especially for first time mothers.

Occasionally, as the due date nears, the cervical plug will be found on the mother's rear end or on the ground. This is a white mucous plug that resembles Elmers® glue. Its function during pregnancy is to keep the cervix, the opening into the uterus, closed. Unfortunately, passage of the cervical plug does not always mean that birthing is imminent. It may be passed several hours to several days before birthing, as the cervix begins to relax.

The majority of births take place during the day, and frequently in the morning, with first time mothers being the ones most apt to deliver in the evening. Normal birthing begins with labor pains which occur several hours before actual parturition. This is called stage one of parturition. This is the initial stage where hormonal changes occur to prepare the mother and baby for the actual birth. Usually, the mother llama has somewhat enlarged teats and wax may plug the tips of the teats. The vulva relaxes and elongates to about three times the normal size. During stage one, she may hum more than normal, not be hungry, may roll, spend more time at the manure pile, lie in kush position, and generally act uneasy or restless. This stage may last two to six hours. Not all llamas give obvious signs of impending parturition. Many mothers are less restless and act fairly normal during stage one and it can occur unnoticed.

Active, hard labor begins in stage two. First, the fetal membranes break. A small amount of fluid is passed during this stage, but often goes unnoticed by the observer. This is the same stage where the "water breaks" in other species. In llamas, however, much of this fluid, plus the main fetal membranes, come out in the afterbirth. When actual hard labor begins, some llamas will lie down and some will remain standing, and many will be up and down. Delivery of the baby occurs normally with the baby's two front feet coming first, followed closely by the nose or vice-versa. Once the head appears, the baby usually slips out quickly, but may take up to 30 minutes. The baby's shoulders are the largest portion to go through the birth canal. With the exception of a rare hip-lock, once the shoulders are out, the baby slides out very easily.

When the shoulders are out, the hips are in the birth canal. This causes the baby's umbilical cord to be shut off as it presses against the mother's pelvic bone. This cuts off the oxygen supply that was the sole source of oxygen during fetal development. Breathing must begin immediately. If the baby's chest is still in the birth canal, the baby cannot expand the chest very well to breathe. This compro-

mises breathing even though the head and neck are all the way out. Therefore, the rest of the baby must be delivered quickly. If the baby seems to be stuck at this point, a firm downward pull on the front legs is all that is needed to deliver the baby. If the baby cannot be pulled out, it is either too large, or in an abnormal position, and you have an emergency. In general, the second stage of labor occurs quickly. Some llamas, especially first time mothers, may take up to two hours normally for the entire second stage.

Now that the baby is finally out, check to make sure it is breathing. Breathing must occur immediately so as not to deprive the brain of oxygen. Well-oxygenated blood will give the mouth and tongue a pink to bright red color. This, in conjunction with regular breathing means that your baby is off to a good start. If you hear or see a lot of mucus or fluid in the baby's nose or mouth, wipe its nose and mouth with a cloth. A bulb syringe will work to help remove excessive nasal or mouth fluid by suction. If the baby still gurgles as it breathes, hold the baby upside-down by the

a.
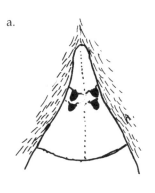

b.
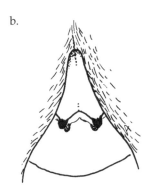

c.
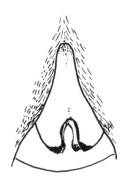

a. Teats of a non-pregnant llama, or a non-nursing mother in early to mid- pregnancy.

b. Teats enlarging about two weeks before birthing, although this varies between llamas.

c. Very full teats. Delivery is imminent. However, much variability exists between llamas in udder/teat appearance. Teats may also look like this just after a baby is weaned or with a very severe case of mastitis involving all four quarters with teats that are hot, red and painful.

Figure 16.4 LLAMA UDDER

hips to allow gravity to bring the fluid to the outside, and then re-wipe its face. If the fluid is excessive, you can swing the baby one to two times upside-down, and let centrifugal force and gravity draw out the fluid from the baby's airways. Additional respiratory stimulation may be obtained by vigorous rubbing of the baby's torso with a towel.

After the baby is breathing well, dip the umbilicus in a disinfectant such as iodine or Nolvasan®. Disinfectant poured into a plastic coffee measurer, a plastic syringe case, or a shot glass works well. Iodine disinfects the umbilicus and helps stop bleeding.

The normal baby is vigorous in rolling and trying to sit up and is usually on its feet and searching where to nurse within one to two hours. Once the baby is dry, you should weigh it and record its weight. If any abnormalities are noted in the baby, or in the birthing process, or if breathing did not occur normally, make sure to have the baby and mother examined by your veterinarian.

Stage three of parturition is the expulsion of the placenta or afterbirth. These membranes are normally passed within 4 to 6 hours following delivery. Someone experienced in llama reproduction should put on disposable gloves and examine the placenta thoroughly for evidence of infection, tears, excessive hemorrhaging, and completeness. If you live in a rural area, take the placenta far away or burn it, or bury it deep in the ground with lime, as the odor will attract wild animals. If the mother llama does not pass the placenta within 12 hours following delivery, do not pull it out yourself. Rather, seek veterinary assistance.

Dystocia

Dystocia means difficult birth. Dystocias are rare, but generally require immediate assistance. They include abnormally positioned babies such that the head and two front feet do not come out first. Such babies cannot fit through the birth canal without getting stuck. Also, a baby may have trouble being born even

Figure 16.5 NORMAL PARTURITION

a. Note elongated vulva

b. Note llama's posture. The fetal sac becomes apparent.

c. Normal fetal presentation

112

d. The baby's two front feet and nose appear.

e. Normal Parturition

if it is in the correct position. The baby may be too large and the shoulders can become stuck in the birth canal. Signs of dystocia may be very obvious. The baby may be stuck or present with just one foot, no feet, or the back feet. These babies usually need assistance to get out.

Another sign of a potential problem is an exceedingly long second stage labor. In general, if your llama is in second stage labor and NOTHING happens within about 1-1/2 hours, you probably have a dystocia and need veterinary assistance. These dystocias are usually due to an abnormally positioned baby that cannot even get started into the birth canal. For example, a baby presenting its back into the birth canal cannot physically get into the canal. After a period of pushing and straining, these mothers may get tired and will typically sit around uncomfortably until help arrives.

Occasionally a prolonged second stage labor is due to the lack of, or weakness of, the mother's uterine contractions. These mothers are unsuccessful in delivery without assistance. If the baby has not started into the birth canal, its umbilicus is still intact and the baby can receive its oxygen from its mother. These babies can live for several hours, so you have time to wait for your veterinarian for assistance. However, if the baby is part-way out and not in a normal position, or is stuck, time is of the essence. This is because the umbilicus is being compressed in the birth canal and delivery must be quick so that the baby can start breathing. With these dystocias, the veterinarian may not be able to arrive fast enough, as it may be only a matter of minutes to save a baby. It may be up to you to manipulate this baby to permit delivery.

If you are inexperienced in llama deliveries and plan to breed llamas, you must become prepared for such emergencies. You must know what is normal parturition to determine what is abnormal. Reading about it helps but is no substitute for observing actual llama births. Since most births are during the day, you can contact a local breeder and be placed on "stand-by" for a phone call when a delivery is expected. It may take several at-

tempts until you actually see a birth, but observing the immediate post-natal period is also important. In addition, you should read about dystocias and talk to many experienced llama breeders and veterinarians regarding what they've learned through their experiences with dystocias. A hands-on perinatal clinic would also provide additional experience and knowledge.

A prepared owner should also have the necessary obstetrical supplies readily available for either normal or abnormal births. The supplies may be kept together in a handy bag or box. Supplies should include several towels for drying the baby, a bulb syringe to aspirate excessive mucus from its nose and mouth, 7% strong iodine for the umbilical cord, umbilical tape for a hemorrhaging umbilical cord, several tubes of sterile lubricant and some plastic obstetrical sleeves for manipulations. Supplies for the mother would include soap, cotton, and tail-wraps.

Even when an owner is well prepared, when it is YOUR llama and the baby's life is at stake, panic sets in, and everything you have read or practiced often exits your mind. For example, often a baby must be pushed back IN before a manipulation can be made. Owners must be reminded of this, as it seems contrary to the goal of getting the baby OUT. Another example of a difficult situation under pressure is the determination of front from rear legs of the baby by feel alone. On the front legs, the fetlock or ankle joint bends in the same direction as the carpus or knee joint. On the back legs, the fetlock joint bends in the opposite direction as the hock joint, which is the next joint above. A cordless portable phone at this point is most helpful, and you can be talked through a dystocia by your veterinarian or be informed that you have enough time to wait for assistance. (Refer to the troubleshooting chart on dystocias for quick reference.) Unfortunately, even the person most experienced with dystocias cannot save all of the babies. Respiration may be so severely compromised and the baby so stressed, that even the most valiant efforts have a sad outcome.

114

Aside from the obvious large or mal-formed baby, the underlying causes of dystocias are rarely apparent. A female that has had normal deliveries may have a dystocia and subsequently return to having normal deliveries. Some reproductive problems are genetic in origin and may result in abnormal fetal development, which can definitely cause a dystocia. Also, the mother may be reproductively inferior. She may have a decreased ability to produce the necessary hormones needed for normal birthing or have an abnormal reproductive tract.

Any health abnormality of the mother may result in a dystocia. This includes infections, inadequate nutrition, poor condition, and obesity. A fat female has a lot of stored fat in the pelvis and abdomen. This decreases the area the baby has in the uterus for development and the size of the pelvic canal available for passage of the baby. These fat mothers are also more sedentary or have too small of an area in which to exercise. In addition to the weight problem, their muscles are not well toned. The uterine contractions are under hormonal control, but the abdominal pushes need healthy and strong abdominal muscles.

Hygiene is of the utmost importance when dealing with a dystocia. Soap, cotton, antiseptic liquid, tail-wrapping gauze, plastic obstetric sleeves, and sterile lubricating jelly should always be available in case a dystocia occurs. Ideally, the mother's tail should be wrapped with gauze, her vulva and surrounding area scrubbed, and your hands scrubbed and sleeved before you begin manipulations. However, once the baby's umbilicus is compressed in the birth canal, you have about three minutes until brain damage occurs to the baby unless breathing begins. In these dire emergencies, just grab the sleeve, lubricate, and proceed. One cannot waste precious time scrubbing. Be sure to mention the lack of hygiene to your veterinarian, as the mother will need antibiotic therapy following such non-sterile techniques.

Note that on the front legs, both knee and ankle joints bend in the same direction. On the hind legs, the hock and ankle joints bend in the opposite direction.

*Figure 16.6 **DIRECTION OF JOINT FLEXATION.***

Troubleshooting Dystocias

1. **The female is over 365 days pregnant. No sign of labor.**

2. **Second stage labor (hard labor) for 1 hour and nothing.**

3. **Second stage of labor for 2 hours and nothing has happened.**

4. **Any Dystocia.**

5. **Normal position, but baby won't come out.**

6. **Hip-lock** (stuck at hips).

7. **Toes of front legs peeking out, but no head appears within 10 minutes.**

1. She may not be pregnant! She may be normally having a long pregnancy. Call veterinarian for consultation.

2. Alert your veterinarian.

3. Likely a dystocia. Call your veterinarian.

4.
   A. Umbilicus in birth canal, work fast!
   B. Umbilicus still in the uterus, time to get your veterinarian.
   C. Always have a follow-up exam for mother and baby.

5. Gently pull downward on the baby's front legs when the mother is having a contraction. Slight rotation of baby or pulling of one front leg at a time may facilitate shoulder passage.

6. If baby's not breathing, stimulate breathing (tickle nose, splash cold water on its face, vigorously rub the baby, or if necessary begin artificial respiration once every 7 seconds) while you rotate baby 90 degrees and help baby out if needed.

7. Umbilicus is not yet occluded. Push feet back in; this may be all that's needed to free up head to normal position. If still stuck, push feet back in and scoop head up into birth canal.

Troubleshooting Dystocias

8. All or part of the head is out, no feet.

8. Probably the feet are close behind. If worried, gently feel inside the vulva, the feet are usually at the top of the opening and can be scooped free towards the vulva. Often in small llamas and alpacas or first time mothers, the vulva is very tight over the baby's forehead. Some lubrication around the head eases delivery. A person with small hands can often perform manipulations easier with these small animals.

9. Head and some neck out, with one or no legs. No change in 10 minutes.

9. Feel to see if one or both legs are folded back. Gently straighten the leg by bending the leg under the chest and pulling the foot forward. Then the baby should come easily.

10. Head and most of the neck out, but one or no legs.

10. Work quickly! Push the baby back IN to the base of the ears or further if needed. Feel for the front legs and gently straighten them one at a time. Then the baby should come easily.

11. Backwards (rear legs presented first).

11. WORKFAST! Deliver the baby backwards but help pull it out to speed delivery or else the baby will suffocate. Rotate the baby 90° at the shoulder level to prevent shoulder-lock.

12. Upside-down presentation.

12. Rotate the fetus. Extend front legs if not already extended. Assist by pulling on the front legs if necessary.

13. Twisted sideways, buttocks trying to come through the birth canal first, mid-back jammed against birth canal. These cause a prolonged second stage of labor with no progress.

13. Alert your veterinarian. These cannot get started into the birth canal, so you have time to seek veterinary assistance. However, uterine torsions which present with prolonged second stage labor and signs of extreme pain, require veterinary diagnosis, and immediate attention.

NOTES

CHAPTER 17:
NEWBORN LLAMAS AND ALPACAS

Introduction

Immediately after birth is a VERY critical time for baby llamas and alpacas. This is a time when tremendous changes occur in the baby to adapt it to life outside of the womb. Prenatally, the fetus gets oxygen and nutrients from its mother. It eliminates waste for its mother to dispose of. It relies passively on mother for a constant source of nutrition, warmth and protection. Once born, the baby must take over all of these functions plus learn to use its legs which it hasn't needed for the last 11 months. These changes for post-natal life are crucial because if any system fails, the llama's or alpaca's health and strength will be severely compromised, even to the point of death.

Immediate Concerns

Once you discover that you have a new llama baby, make sure it is breathing. A newborn baby that is breathing normally should have pink to red mucous membranes and tongue. If your baby is not breathing or its tongue and gums are pale white or blue, make sure the airways are clear. Wipe any mucus from the nose and mouth. A bulb syringe works well to draw the mucus from the nose and mouth. Immediately stimulate breathing by tickling the baby's nostrils or splashing some cold water on its face or rubbing its torso vigorously with a towel. If this does not work, artificial respiration should be done. If no res-

pirator is available, exhale into the baby's nostril either directly or through a small tube. The other nostril and mouth must be covered or held closed. Repeat every seven seconds until the baby begins to breathe on its own.[1]

A congested sound to breathing usually indicates fluid in the airways and can be remedied as described in Chapter 16. If your baby continues to have difficulty breathing and only breathes with its mouth open, consult with your veterinarian about the possibility of choanal atresia. (See Chapter 13.)

If the baby is breathing well, next dip the umbilicus in disinfectant such as iodine (2 to 7%) or Nolvasan® solution (0.5%).[2] Dunking the hanging piece of umbilicus into a shot glass filled with disinfectant works well. The umbilicus is a very common entry site for infection in a newborn. Both Nolvasan® and iodine act as excellent antiseptics. Iodine also helps stop bleeding and promotes drying of the umbilicus. When the umbilicus snaps at the time of birth, it will dribble blood for a couple of minutes. If it hemorrhages profusely, direct pressure can be applied to the umbilical cord to stop the bleeding. Alternatively, umbilical tape or a new shoe lace dipped in iodine can be tied about 1/2 inch from the baby's belly. The tie should be removed several hours later. It should be pointed out that the other end of the umbilicus is attached to the placenta, which most mothers have not passed at this

119

point. Thus, she will bleed a little too from the vulva when the cord breaks. It should not be a cause for concern, unless it is profuse or continuous.

If the baby is in a cold area, dry it off. Towels and hair dryers work well. If the baby was born in a cold and muddy spot, you can gently carry it to a cleaner and warmer area. These precautions will decrease the chance of hypothermia. Once dry, the baby should be weighed so that its progress can be monitored. Normal babies can weigh 18 to 40 pounds, with the average being 27 to 30. Alpaca crias average 16 to 20 pounds.

The above described dipping, drying off, and weighing should be done quicky to allow the baby and its mother to bond as soon as possible. Humans should not interfere with the mother to cria bonding. Too much human interference at this crucial time can slow the bonding and nursing. A good compromise is to observe from a distance if they are outside or put a small peephole in the barn if they are inside and quietly observe the pair. Closed-circuit television would be ideal, but is quite expensive. Normal babies try to stand up almost immediately and the human temptation is to help. However, the attempts actually strengthen the baby's muscles and tendons. Most are on their feet within two hours and have nursed within six hours.

Nursing

The baby is born with very little protection against disease. Its immune system, which fights off disease, is still immature at this stage. Until its own immune system matures, it relies completely on antibodies (infection fighting proteins) from its mother's milk. These antibodies are found only in her first milk which is called colostrum. This is why it is important to confirm that nursing has occurred. Without colostrum, a microorganism is bound to get the upper hand in infection of the new baby. This is the reason why the mother should be current on all annual vaccinations which boosts the level of antibodies in her system and passes these cruicial antibodies on to the colostrum. If there is doubt as to whether the baby has nursed, your veterinarian can run a simple blood test on the baby to reassure you

(a Radial Immunodiffusion (RID) laboratory test or a sodium sulfite test, for example.)

If your baby has not stood in two hours or nursed in 6 to 8 hours, it is time to intervene. First, stick your finger in the baby's mouth to check for a sucking reflex. If it sucks, this is good. If not, call your veterinarian now! A baby without a sucking reflex will need to be fed through a stomach tube. It may also need critical care and supervision.

If the baby has a normal sucking reflex, next check the mother's teats. Remove any waxy plugs at the end of the nipples. Make sure she has colostrum. If she does not, phone the veterinarian to request medication to aid in milk let-down. If drugs such as oxytocin don't work, goat colostrum or the rarely available llama colostrum should be fed to the baby.

If the baby can stand, has a sucking reflex, and the mother has milk, all that may be needed is to point the baby in the right direction. Rubbing the baby and lifting its tail over its back, a normal sign of hunger and desire to nurse, tend to stimulate nursing. If this effort is unsuccessful, you can give the newborn milk from a bottle. You can milk the mother if she is gentle, but this is easier said than done, and the quantitiy obtained is often insufficient. Otherwise, use goat colostrum. The baby needs 5 to 10% of its body weight in colostrum, divided into three to six feedings within the first day or preferably within the first eight hours of life. Thus a thirty-pound baby needs one and one half to three pints. (See Chapter 4 for more information).

If it has been longer than 24 hours since birth, and your baby has not received any colostrum, obtain veterinary assistance. A plasma transfusion may be needed, as well as tetanus and enterotoxemia anti-toxins.

Premature Baby

In general, premature babies are underweight. Llama crias may weigh less than 18 pounds and alpaca crias less than 9 pounds.[3] Other signs of prematurity include the absence of erupted incisors, fine fiber, the retention of the protective caps covering the toes, a poor sucking reflex, weakness, droopy ears, and difficult breathing. Premature births can be due to fetal abnormalities. Also, infections and stress for the mother can induce an early labor. Stresses include long trips, environmental extremes, and illness.

One of the last fetal developments is surfactant production in the lungs. Surfactant serves to decrease the surface tension in the microstructure of the lungs and therefore prevents collapse of the tiny airways. Premature babies often do not have adequate surfactant, and this makes breathing difficult to impossible for them. The baby's tongue and gums may appear pale to blue. The baby should be given oxygen if available, until it pinks up. Prolonged, straight oxygen, however, should be avoided. If the baby is not breathing at birth, artificial respiration should begin immediately.[4]

Post-Natal Concerns

Once your baby has stood and nursed, it should be observed to make sure it is able to urinate and defecate normally. The meconium is the first feces passed. It is a brown-black, hard, ball-shaped feces. The fetus normally ingests fluid while in the womb. As this is digested, the waste is stored in the baby's colon and rectum as the meconium. Since meconium is so hard, it is sometimes difficult for the baby to pass it. A cria with a retained meconium will show excessive straining, raised tail, arched back, and may show little desire to nurse and may roll in discomfort. The treatment is an enema. A Fleets® enema, which is available at drug stores, with a well-lubricated tip (use K.Y. jelly® or Vaseline®) or a mild soapy warm water solution works well. Use extreme gentleness when administering an enema and do not force the enema tip in any further than through the anus. Some crias may need a second enema, if they are still straining. However, repeated enemas will cause rectal irritation, which causes signs similar to the original constipation! Most babies will begin to defecate immediately following an enema, but some may take up to twenty minutes.

The opposite problem to watch for is diarrhea in newborns. The serious complication here is dehydration. With small fluid reserves, a baby rapidly loses fluids in the diarrhea, and this creates an extreme emergency situation.

If the baby is otherwise normal, and is not dehydrated, the diarrhea may be due to excessively rich milk. In this case, a little oral Kaopectate® or Pepto-Bismol® given slowly with a needleless syringe may be all that is needed. However, if the diarrhea is accompanied by a weak, sluggish, or feverish baby, veterinary attention is needed. These babies will need fluid and antibiotic therapy.

Baby llamas also may have wobbly or crooked legs, dropped ankles, or knock-knees. Most of these are just due to weak ligaments and from positioning in the womb. Time will take care of the majority of these problems. Unless the defect is so severe that the cria cannot walk, wait two to four weeks before being overly critical of your baby's legs.

A few baby llamas may be born with floppy ears. This may be the entire ear, or only the tip. The rigidity of the erect ear is due to a cartilage framework inside the ear. Usually, the cartilage will strengthen on its own and the ear will soon be erect. Ears that show no signs of improvement in three to four days may need some support. If external support is provided, it must be done with the utmost care. Tape should be used for external support, but if taped too tight, especially when the weather is cold, the blood supply will be cut off to the ear and result in the ear dying. If the support is too heavy, it may increase the likelihood of a crease or bend in the ear. A moleskin used for shoe inserts can be cut out to fit just inside of the external ear. Alternatively, a small gauze roll can be placed in the ear and held with two narrow strips of tape. This provides some support without being too heavy. Remember, these ear splints can cause damage, if done incorrectly or used when they are not needed. Furthermore, a baby born during cold weather with normal ears which subsequently begin to bend at the tip could indicate frozen ear tips. These tips will probably die and eventually fall off.

Normal baby llamas' daytime activities include a series of several short naps. Between naps they nurse, explore their new world, and may play a little. Normally, they have a very active play period in the early morning and evening. Now you can enjoy watching your baby llama play and grow.

Troubleshooting Newborn Problems

1. **Baby not breathing or pale.**

1.
   A. Open airways, wipe nose and mouth.
   B. Tickle nose or splash cold water on baby.
   C. Hang upside down briefly.
   D. Perform artificial respiration once every seven seconds.
   E. Obtain veterinary assistance.

2. **Umbilicus hemorrhaging.**

2. Apply direct pressure to umbilicus or tie umbilical tape 1/2 inch from belly. Dip in 7% iodine.

3. **No nursing after 6 hours.**

3.
   A. Check baby for suck reflex. If none, call veterinarian.
   B. Check mother for presence of milk. If none, call veterinarian.
   C. Unplug nipples.
   D. Guide baby to teats.
   E. If unsuccessful, try bottle feeding with llama or goat colostrum or feeding via a stomach tube.
   F. Obtain veterinary assistance.

4. **Baby lethargic and straining to defecate, or no bowel movement in 12 hours.**

4. Give an enema with a well lubricated tip.

5. **Diarrhea.**

5. If depressed, feverish, dehydrated or not nursing, call veterinarian. If acting normal, no fever, no dehydration, observe it closely and give Kaopectate® orally for relief if diarrhea persists.

CHAPTER 18:
MASTITIS

Introduction

Llamas normally have four teats or sections to their mammary glands. The nipples of the four teats can be seen underneath the llama, between the hind legs. Some llamas have more nipples, but usually the extra nipples are nonfunctional. The mammary glands produce the colostrum and milk that are vital to the raising of their offspring. Mastitis is an infection of the mammary glands. It is not common in llamas and, therefore, some of the following information has been extracted from cattle research.

With llama mastitis, one or more of the teats may be affected. Signs of mastitis vary with the severity of the infection. The disease may be mild and not apparent without performing diagnostic laboratory tests. The other extreme is an extremely ill, depressed, and feverish llama. Llamas seem to be most susceptible to mastitis just after parturition and just after weaning their baby. The llama may be predisposed to mastitis if she has teat injuries or sores, or if she naps in a dirty and muddy area.

Figure18.1 MASTITIS, LEFT REAR TEAT.

Subclinical Mastitis

Subclinical mastitis is a slight infection that shows no outward signs. The llama acts normal and her milk appears normal. It is likely that this is the most common mastitis in llamas and it exists unnoticed. Perhaps these are the females that raise slow growing babies year after year. A milk infection of the mammary gland will cause a decrease in milk production and abnormal milk composition. Her baby often will not nurse on the teat with mastitis when only one is affected. Milk samples should be evaluated by your veterinarian if a baby is a "poor doer" or if the baby is not nursing on all four teats regularly.

Subacute Mastitis

Subacute mastitis is also a mild form of mastitis. It is likely the next most common form of mastitis. Only the teats are affected, and the llama as a whole is acting normal. The affected teat may be swollen and slightly reddened. Usually, the baby won't nurse the affected teat due to a bad taste in the milk. The milk may also have a slight change in consistency, either watery or thickened.

Acute Mastitis

Acute mastitis is more severe than the above forms. The llama may be depressed, have a fever, and be off her feed. The affected teat is very warm to the touch, swollen, and red. The milk is blood-tinged and the consistency is abnormal. The milk may be very thick with yellow clumps, or it may be watery with clumps. The milk often smells bad, too. If the baby is nursing on the affected teat, it may also be ill. Diarrhea is the primary sign in the cria.

Peracute Mastitis

Finally, mastitis can be peracute. This is a sudden and serious illness that is usually seen within a few days of parturition.[1] These llamas are extremely sick. The teat is red, hot, swollen, and painful. The milk is very abnormal in color and consistency and contains blood. In addition, the llama does not want to eat, is extremely depressed, feverish, and is usually lying down. These cases are emergencies!

Diagnosis and Treatment

Diagnosis is based on observing the above mentioned signs and examining a milk sample. The milk can be observed for changes in color and consistency. Abnormal milk usually contains blood. However, visual examination alone will not identify the subclinical mastitis or some of the subacute mastitis cases, as the milk may appear normal to the naked eye. Thus, tests for the presence of microorganisms or abnormal composition of the milk should be used. When the mammary gland becomes infected, the number of cells (mostly white blood cells) greatly increases. Tests are available to approximate the number of cells in milk samples. The California Mastitis Test is an example of one of these tests. The California Mastitis Test (C.M.T.) is a quick screening test used for dairy cows. It also seems to have consistent results in llamas. It is a cheap and fast test available commercially. The test kit contains a solution with a purple color indicator that causes the purple milk mixture to turn a very deep purple in most cases of mastitis. Also, the sample will turn to a thick gel as the number of cells in the milk increases, thus indicating mastitis. To use the C.M.T., test each teat separately. Squirt a few drops of fresh milk from each of the four teats into the four separate compartments of the paddle provided in the kits (or into four separate containers first to avoid spilling). Then, add equal amounts of the purple test solution. Positive samples will form a gel. In cattle, a score of negative, trace, one, two, and three are used to describe the amount of gel that forms, with three being a very thick gel indicating very abnormal milk. Before using the C.M.T. on a suspected case of mastitis, evaluate a normal llama that is exhibiting no signs of mastitis. This enables you to interpret the appearance of a negative to trace result.

Finally, a culture and sensitivity should be run on the milk of any llama suspected of mastitis. A culture is a laboratory test to isolate and grow harmful microorganisms from the milk sample. The sensitivity test determines which antibiotic drugs work the best at killing the harmful microorganisms present. However, it takes a few days to run a culture and sensitivity, so antibiotics such as penicillin are often started earlier, until the antibiotic of choice is revealed.

The antibiotics are the most important therapy. However, frequent milking of the affected teat also helps rid the llama of mastitis, by flushing out the toxins or poisons from the udder. In addition, cleanliness is important, as you don't want your dirty fingers to re-infect the teat. Wash your hands, and cleanse the udder before and after treatment.

Sometimes, teat infusions help clear up mastitis. These are antibiotics squirted up into the hole of the teat. Many intramammary infusions are available commercially for dairy cows. Due to the small size of the llama teat opening, the cow infusion tips are too large to insert into the llama teat. Tom cat urinary catheters work well in llamas for administering these preparations. Since double canals are associated with each teat, both canals must be infused.[2] To infuse the teat, the llama must be properly restrained, cooperative, and the infusing must be done very gently. It is advisable to have your veterinarian do the infusing or show you the technique, because if done incorrectly, it could easily cause further damage to the teat.

If the llama has mastitis signs and is ill (depressed, not eating, feverish), begin by contacting your veterinarian. In addition to the above localized teat therapy, the llama will need intensive supportive therapy such as intravenous fluids and systemic antibiotics. If her baby is nursing from infected teats, or the mother is too ill to allow the baby to nurse, temporarily remove the baby from access to the mother until the mother is better.

Prevention

Preventative measures are always preferred for animals that tend to develop mastitis. General cleanliness can decrease the chance of mastitis, with extra cleaning efforts made around the time of birthing. The pen that the female lies in and gives birth in should be kept clean and dry. Obviously, lying in a muddy, feces packed area would increase the chance of mastitis. Also, dipping the female's teats when the baby is removed for weaning will decrease the possibility of mastitis. Dipping the teats destroys the harmful microorganisms at the end of the teat and therefore decreases the chance of them migrating up the teat canal. It also promotes healing of any existing sores and prevents the development of new sores. Numerous teat dips for cattle are available. Even Betadine® solution may be used as a dip. The dipping is just as the name implies. Pour the dip into a shallow container (clean coffee measurers work well). Dip all teats. With these preventative measures, and early detection and treatment, mastitis and its complications can be kept to a minimum in your herd.

NOTES

CHAPTER 19:
THE AGING LLAMA OR ALPACA

Introduction

With the superb care owners are providing for their llamas and alpacas, many are living very long lives. Good nutrition and husbandry, preventative medicines and skilled veterinary services combine to help llamas and alpacas live well into or past their second decade. However, many changes occur with the inevitable aging process. Aging is an irreversible and cumulative process which is genetically programmed in each individual. This genetic information coupled with environmental influences determines the rate of aging. Thus, much individual variation occurs in regards to rates and manifestations of aging.

Musculoskeletal Changes

Aging brings decreased strength in muscle and bone tissues. This is combined with the cumulative affect of years of wear and tear on joints, decreased muscle mass, plus residual affects of a lifetime of various injuries. Older animals also have a decreased ability to recover from new injuries. Thus some older llamas and alpacas experience pain and stiffness when moving. Pain from arthritis can occur in any joint but seems to be relatively common in the neck and knees. There will be days when a pain reliever would certainly help an arthritic animal. However, do not overuse pain relievers and make sure to consult with your veterinarian as to the best drug to use, (of course none are labeled for usage in camelids) an

appropriate dosage and how often you could administer such a drug. The reason you should use caution on what seem to be innocuous drugs is that they directly irritate the wall of the stomach and can result in an ulcer in species that are already prone to ulcer development. Other very safe means of helping your stiff and arthritic animal include using a heating pad held on stiff joints for a half hour, massaging sore muscles, or applying topical liniments to sore areas (not to open wounds). Provide the animal with a rest area and warm housing during cold weather for comfort. Alternative medicine, such as acupuncture, has had great results in many llamas with chronic pain. More and more veterinarians are becoming trained and certified in acupuncture techniques. In fact, some veterinarians with specialized training, limit their practice to only acupuncture patients.

Temperature Extremes

With senescence (aging), the animal decreases its ability to adapt to changes in the environment. Some aged animals are too fat and can have life-threatening problems with hot environmental temperatures (as can younger animals). It is imperative to shear some fiber on heavy-fibered animals, avoid stressing them during particularly hot days, and provide shade and fresh clean water at all times. Additionally, decrease caloric intake to help fat animals lose those extra pounds, as this extra weight not only contributes to hyperthermia, but leads to problems in major body organs. (Please refer to Chapter 12 for more information on heat-related problems.)

Very often the aged llama or alpaca is thin and frail, rather than fat. This can be due to their decreased ability to compete for food, muscle atrophy (wasting) or poor dentition. The thin older animal has very little natural fat insulation and may also have thin, short, poor quality fiber. Obviously, this animal can become chilled rapidly in cold weather, especially if it is wet and exposed to wind. As the animal attempts to warm itself through shivering, energy reserves can be depleted. If the animal cannot successfully restore normal body temperature, hypothermia results, which, if severe, can lead to a coma and could result in death. (See Chapter 12 for more information on hypothermia.) Therefore, special consideration should be taken to make sure older animals have a shelter from wind, and in particularly frigid areas, access to heated rooms. If a herd shares a shelter, make sure the older animal is not ousted from using the shelter by more dominant animals. A blanket or neck-warmer can be used for extra warmth, but make sure that it never remains on the animal if it becomes wet. Lastly, ensure a high quality diet to help the aged llama or alpaca maintain (or gain) weight and provide extra feed necessary for warmth.

Vision and Hearing

As with humans, aging brings about degenerative changes in sensory structures such as eyes and ears in llamas and alpacas. It is amazingly difficult to detect changes in vision and hearing in your older animals, especially if they have lived in the same environment for years. Usually age-related changes occur slowly, over many years, and the animals learn to adapt quite well to sensory deficits.

Eye changes can be detected by ophthalmic examination by your veterinarian, provided you have a dark area (like a dark stall) for the exam and the animal is cooperative by holding its head still. Although the exam is painless, some animals are annoyed at the light instrument shining in their eyes and the re-

peated attempts to restrain their head. Auditory function tests are much more subjective and involve simple observation of reactions to various noises as well as cursory ear exams. Auditory dysfunction is often difficult to detect.

Care of a visually or auditorily impaired animal involves few changes except being aware of their disability. Avoiding startling the animal by talking to blind animals to notify them of your presence and approach deaf animals so that they can see you. Minimize changes in their daily routine and environment to avoid unnecessary stress and confusion.

Digestion and Nutrition

Normal digestion and nutrient absorption can occur in old age but some problems may arise. The most common digestive problem in senescence is a dental problem. Older llamas and alpacas commonly have missing teeth, infected and painful teeth, or chipped teeth resulting in inefficient chewing and rechewing of feed. Without proper chewing, which is a very important step in the breakdown of food, many nutrients pass through the digestive tract without absorption into the

Figure 19.1 OLDER LLAMAS "OVER AT THE KNEE" ON FRONT LEGS, MORE PRONOUNCED SHOULDER HUMP AND SWAY BACK.

bloodstream. Evidence of dental problems includes lumps or swellings on the upper or lower jaw, noticeable broken teeth, particularly malodorous breath, dropped wads or partially chewed feed or regurgitated feed and noticeable undigested feed present in the feces. It is important that your older llamas and alpacas have annual exams that include dental inspection. If dental problems exist, treatment may be possible. However, if teeth are missing or broken, the animal should be placed on an easy-to-chew diet such as chopped feeds or softened (with warm water) feeds.

The natural aging process brings about some changes that don't directly cause problems but can make your llama or alpaca more susceptible to digestive upsets. The lining of the digestive tract becomes more prone to damage which could lead to ulcers. (See Chapter 15 for more information on ulcers.) This ulcer potential could be compounded with chronic usage of pain relievers such as phenylbutazone and aspirin.

Lastly, the motility (muscular movement of the digestive tract) decreases a small amount with aging and thus digestive contents pass through the animal more slowly. This could predispose to intestinal obstructions and colic especially if the older llama or alpaca isn't drinking sufficient amounts of water due to painful teeth or difficult access to water sources. Good quality nutrition - including easy access to fresh water and a salt/mineral mix, good husbandry habits - such as regular deworming, plus close observation will minimize digestive problems in your herd.

Immunity

With aging, the immune system becomes less effective at combatting infection and detecting cancerous cells. As a result, older animals are more susceptible to infectious diseases and to complications of such diseases. For example, when a respiratory infection spreads through your herd, keep a very close watch on the very young and the very old populations. Complications, such as pneumonia, are more likely in these populations. It is

important to keep vaccinations up to date on older animals as well. (Refer to Chapter 5 for appropriate vaccination program in your area.) You might think that the older animal has been vaccinated so many times during its life that it shouldn't need additional boosters. However, prior immunity may fade with age and immune responses generally decrease, and therefore, regular vaccinations in older animals are crucial.

Just as in humans, most types of cancer are more common in older animals, partially due to immune surveillance not detecting abnormal cells until they are too numerous to combat. Any change in your animal's eating, urinating or defecating habits, any abnormal discharge, any obvious lump or non-healing sore could be an indication of cancer. Additionally, yearly veterinary examinations serve as early detection for cancer so that appropriate treatment can begin rapidly.

General Organ Dysfunction

Age-related problems can manifest in different major organs, because the aging process affects all bodily systems. From impaired vision to diminution of kidney cells to liver damage, aging inevitably leads to some degree of functional impairment of organs. Some organ damage is permanent and irreversible, while some ailments, such as cardiac arrhythmias, can be treated successfully. Although total cures are rare, some problems can be significantly slowed in their progress. Thus, it is important to know the status of major organ systems in the aged animal to detect problems early in their progression. It is imperative that your older llamas or alpacas have yearly complete veterinary exams to assess all major organs, such as eyes, ears, heart, lungs, lymph nodes, bones and joints, muscles and mouth - including teeth. (What is meant by your "older llamas or alpacas"? Older is 15 years and over, but individuals certainly show effects of their age at different rates. A 12 year old llama may be showing signs of old age whereas a 20 year old llama may act like a 10 year old. Regard your llamas and alpacas as individuals.)

Additionally, routine blood tests including blood cell counts and serum chemistries offer vast amounts of information regarding the functioning of internal organs. These blood tests also offer your veterinarian invaluable information when choosing the best medications or anesthetics when the need arises. If all blood tests are normal, these values provide a good baseline comparison for subsequent blood tests. The interval between screening blood tests depends on the overall health status of your animal and the initial results from these tests. Minimal recommended intervals range from every six months for those having organ problems or ill thrift to annually for older healthy llamas and alpacas.

With the considerations listed above, your llamas' or alpacas' older years can be good quality and comfortable years. Without obvious weakness or musculoskeletal problems, older pack llamas can still go on short pack trips. As endurance decreases, the llama should not be forced to go on long trips. Public relations llamas and alpacas can still be used to educate and bring joy to observers as long as the trips are short enough not to unduly cause stress. Some llamas and alpacas can continue to reproduce as they age, but males may tire more easily, have decreased fertility, or have sufficient musculoskeletal problems to make breeding a painful ordeal. Females that are thin with poor body condition should

not be expected to expend calories for a fetus. Additionally, a female's uterus may have remnant scar tissue of multiple births, and\or infection. An infection may or may not be observable by anyone and usually is caused by bacteria entering from the vagina especially at the time of mating or at delivery. Obviously, the older female has had many opportunities for uterine damage and thus may not be a good candidate for breeding. The decision of when to stop using an animal for breeding can be determined with the help of your veterinarian's input upon examination.

Physiological responses to aging are so varied that it is important to consider each of your animals as individuals. While many animals can be quite active in their older years with few problems, unfortunately, the difficult decision of euthanasia may arise for some. The animal may become very ill or so crippled that euthanasia becomes a humane decision. Although this hard decision ultimately belongs to the owner, input from veterinary exams, blood tests, responses to medications, appetite, and general attitude of the animal provides valuable information for making the best decision at the appropriate time. Many older animals do quite well with aging and just need an extra dose of "TLC". With particular attention to certain health issues and careful management, your older animal friends could have many good quality years ahead of them.

Figure 19.2 *OLDER LLAMA UNABLE OR UNWILLING TO RAISE NECK DUE TO PAIN AND STIFFNESS*

130

CHAPTER 20:
HOW TO GIVE INJECTIONS

Introduction

In general, all injections or blood samples should be done by your veterinarian. If done incorrectly by an inexperienced person, the animal could die. Occasionally, a llama will have a hypersensitivity (anaphylactic) reaction to a medication and your veterinarian has the life-saving drugs and knowledge available, but untrained owners could not obtain help fast enough to help or save this llama in trouble. Improper technique could damage internal structures such as nerves, muscles, blood vessels, or bones. Also, life threatening infections could be caused by poor technique, as you are invading the llama's internal tissues by sticking it with a needle. Drawing a blood sample could introduce bacteria directly into the bloodstream! However, there are instances when you need to give injectable medications due to the frequency of needed injections or remoteness of your location. This likelihood should be discussed in detail with your veterinarian before the situation arises in order for you to learn proper technique. If you do not have this personalized training or do not feel comfortable injecting or drawing blood, DO NOT DO IT, as you may cause more harm than good.

General Rules of Giving Injections

1. The animal must be restrained.
2. Swab the injection site with an alcohol swab.
3. Do not touch the needle, except by the hub, so that it remains sterile.
4. Fill the syringe with the amount of needed medicine and then remove the air bubbles from the syringe.
5. After inserting the needle into the injection site, always aspirate the plunger before giving the drug. This means pulling back on the syringe plunger. If blood comes back into the syringe, do not give the drug. Remove the needle and start the injection process again in a different location. Some drugs are lethal if accidentally given into a blood vessel. If no blood draws back into the syringe, administer the drug.
6. Push the plunger with steady pressure to inject the medicine. Do not redirect the needle without first checking for blood by aspiration.
7. Always use a new or sterile needle for each injection.

Figure 20.1   PARTS OF A NEEDLE AND SYRINGE

Figure 20.2 FILLING A SYRINGE

a.

a. Handle the needle by its plastic protector. Remove the cover over the hub and put the needle on the syringe. Do not touch the needle or syringe nozzle. Twist the needle on tightly.

b.

b. Without removing plastic needle protector, pull the plunger back until the edge of the rubber washer is at the volume mark for the quantity of medication needed; in this case, 2-½ cc.

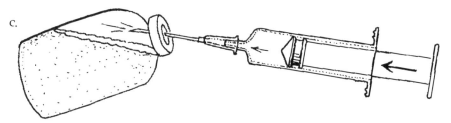

c.

c. Clean the rubber top of the medication bottle with an alcohol swab and let it dry. Remove the needle cover, insert the needle through the rubber portion of the bottle-top, and inject the air to equalize the pressure of the medicine you're removing.

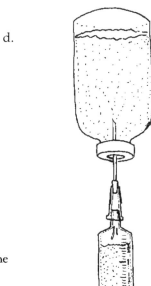

d.

d. Without removing the needle, hold the bottle upside down, and pull the plunger back until the desired quantity of medication is in the syringe.

132

e. Withdraw the needle and syringe from the medication bottle. While holding the syringe vertically, tap the syringe firmly with your finger to get air bubbles to rise to the top. Then push gently up on the plunger to expel air. Recheck the volume careful at the top edge of the rubber washer. If the volume is less than required, get more medicine as in "d" above.

e.

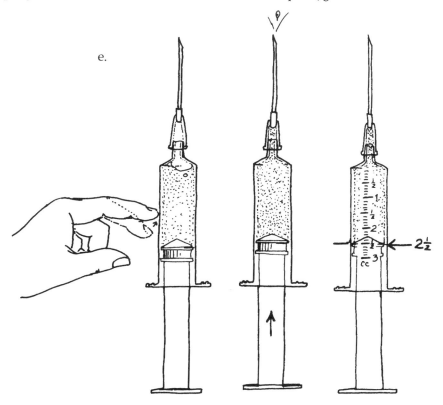

Intramuscular Injections (IM)

1. Any large muscle mass will work.
2. Use a 22 to 16 gauge needle. The thicker the drug, the larger the needle, and the smaller the gauge number. Use a 1-1/2 inch length needle.
3. Avoid the sciatic nerve. (See Figure 19.3a.)
4. Avoid the neck region.
5. The needle should be inserted all the way in to make sure you are in a muscle. (See Figure 19.4.)
6. If frequent injections are needed, rotate the muscles used to avoid soreness or better yet, switch to SQ injections.

Subcutaneous Injections (SQ)

1. These injections go in the layer between the skin and the muscles below. (Figure 19.3.)
2. Use a 22 to 16 gauge needle by 1 to 1-1/2 inches long.
3. The needles should be directed through the skin. With practice, you can feel when you've gone through the skin. This is where you inject the drug. If you penetrate any deeper, you'll be in the muscle.
4. By grabbing the hair near the site of injection with your left hand (if you're right handed), the skin can be slightly elevated. This makes it very easy to feel when you've passed through the entire thickness of the skin. (See Figure 19.5.)
5. If there is a lot of resistance when pushing the plunger, you are probably still in the skin, not below it.

Figure 20.3 INJECTION SITES

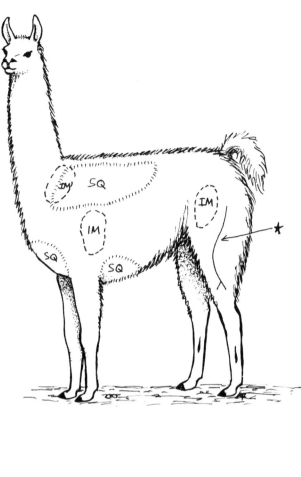

Figure 19.3a
IM=Sites for intramuscular injections.
SQ=Sites for subcutaneous injections.
* = Sciatic nerve.

Llama's
Neck

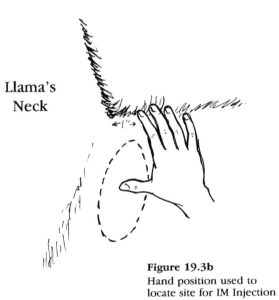

Figure 19.3b
Hand position used to
locate site for IM Injection
on shoulder

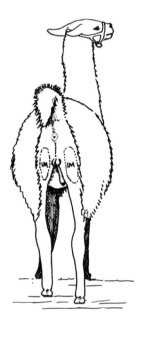

Figure 19.3c

Figure 20.4 INTRAMUSCULAR INJECTION

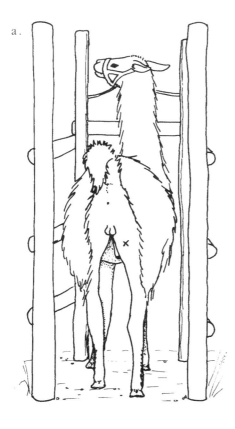

a.

a. Select the site. Choose a site that is well-muscled and where you cannot feel bone. Clean the site with an alcohol swab. The short-fiber area in the hind leg is shown here for purposes of illustration.

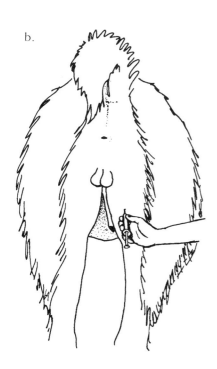

b.

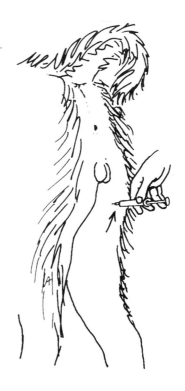

c.

c. BEFORE INJECTING, pull slightly on the plunger. Look for blood entering the syringe at the needle hub. (See arrow.) If NO blood is visible, inject the medication slowly, then remove the needle and syringe. If blood is visible, pull the syringe and needle out and try again.

b. Put the needle straight in, all the way.

135

Figure 20.5   SUBCUTANEOUS INJECTIONS.

a.

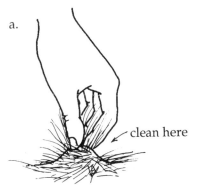

clean here

a. Choose the site. You need a place where the skin is fairly loose, so that you can pinch a fold of skin and pull it slightly away from the body. Clean the skin thoroughly.

c.

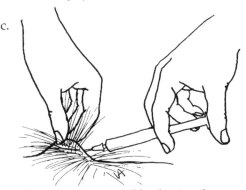

c. If you cannot see any blood, inject the medication slowly into the space between the lifted fold of skin and the body muscle. Withdraw the needle and syringe. Depending on the quantity used, you may be able to feel the medication as a soft bump just under the skin.

Collecting Blood Samples

1. This should preferably be done only by a veterinarian or a person with medical training.

2. An untrained person should never attempt to get a blood sample from the jugular vein in the neck. There are too many vital structures that could be accidentally damaged by an untrained person such as the esophagus, nerves, windpipe, and carotid artery.

3. With practice, small amounts of blood such as those needed for progesterone assays, can be obtained from an ear or tail vein.

b.

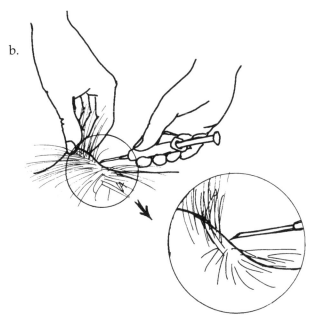

b. With the bevel up, and the syringe more or less parallel to the body, push the needle through the skin. Check to make sure the tip of the needle does not go all the way through and out the other side of the fold of skin! As in step 19.4c, pull back on the plunger to check for blood.

Ear Vein

1. The veins are readily seen on the ear. The trick to this method is proper restraint, so the animal's head is not moving.

2. Cleanse the ear with alcohol.

3. Tie a piece of rubber tubing at the base of the ear to restrict return flow of venous blood.

4. The ear veins are now distended and can be easily seen on the outside of the ear.

5. Gently insert a 20 to 21 gauge by 3/4 to 1 inch needle and pull back on the plunger to obtain blood.

6. After obtaining the sample, immediately untie the rubber tubing. Ideally, a helper can do this just as you finish drawing the sample.

7. After you pull the needle out, put some pressure on the hole in the ear to prevent excessive bleeding.

136

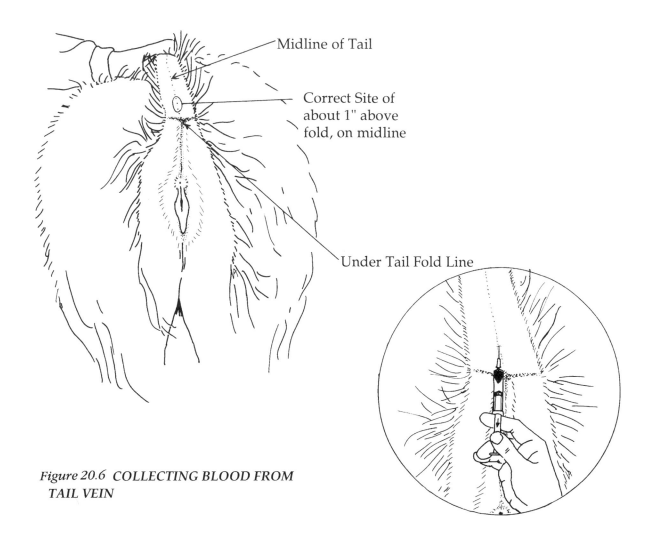

Midline of Tail

Correct Site of about 1" above fold, on midline

Under Tail Fold Line

Figure 20.6 COLLECTING BLOOD FROM TAIL VEIN

8. The blood should be transferred into the shipping container in less than three minutes or it may clot up in the syringe and needle. Ideally, you can hand the syringe with blood to a helper to transfer into the container, while you hold pressure on the ear to stop the bleeding.

9. Once no blood oozes from the site, you are finished.

Tail Vein

1. The tail vein runs down the center of the underside of the tail. Therefore, the landmark is midline. You usually cannot see the actual vein. Some llamas have a very small tail vein and the blood may come slowly.

2. Restrain the llama.

3. Lift the tail up and cleanse the area with alcohol.

4. Continue to hold the tail up with your left hand, while you gently insert a 20 gauge by 1 inch needle on about a 6cc. syringe. Insert this needle in the center about 1/2 to 1 inch away from the attachment of the tail to the body.

5. Insert the needle straight in and the needle will stop when you hit the tail bone. Now aspirate on the syringe. You will get blood either here or perhaps a slight distance out from the bone.

6. If the blood is coming very slowly, you can relax the tail hold a little to encourage blood flow.

7. After obtaining the blood, hand the sample to a helper to put into the shipping container.

8. Hold pressure on the puncture site until bleeding ceases.

137

NOTES

CHAPTER 21:
PASSING A STOMACH TUBE

Introduction

Sick animals and problem newborns often need medicine, food, milk or some other fluid administered with a stomach tube. This procedure ideally should be demonstrated to you by someone with experience. The procedure is not difficult, but it is imperative that it be done correctly. Done incorrectly, you could kill your animal. Prepare ahead of time for emergencies by having the equipment described below readily available.

Baby Llamas and Alpacas

Newborn llamas that cannot or will not nurse from their mothers or from a bottle need their nutrition (colostrum, milk) via a stomach tube. Necessary supplies include a flexible rubber tube. A puppy feeding tube works well. The tube diameter should be a #10-24 French catheter (24 French for alpacas) with the average baby using a tube about the diameter of a pencil. The tube should be about 18 inches long. Also, make sure the tube has no rough edges.

The baby should be put in the kush position with the head upright. The person passing the tube can easily kneel straddled over the baby's back without actually sitting on the baby. The tip of the tube should be lubricated with lubricating jelly. The tube is inserted over the base of the tongue, taking care to avoid the sharp teeth. Pass the tube gently down the throat and into the esophagus. The esophagus is the tube that connects the throat with the stomach. As you pass the stomach tube down the esophagus in the neck, you should be able to feel the tube pass down the neck. The stomach tube can be felt in the esophagus just beside the rigid trachea.

The stomach tube must NOT be passed into the trachea. The trachea, or windpipe, is comprised of cartilage rings, making it a fairly rigid tube. The esophagus lies a little to the left of the trachea and is a soft collapsible tube. If the stomach tube you are passing is in the esophagus, you can feel the rigid trachea in the neck with your fingers, and will feel the stomach tube sliding next to the trachea. (See Figure 20.1.) If there is doubt, slide the stomach tube up and down and feel the neck for its presence. If the tube erroneously went down the trachea, it would be a tube within a rigid tube, the trachea, and it could not be felt passing down the neck.

When giving milk or colostrum to a cria, the tube should only descend to just before the opening of the chest cavity, where the neck meets the chest. This is because the digestion of milk is different from what the digestion will be when the baby is on solid feed. The milk given into the esophagus bypasses the rumination compartment for more complete absorption (for colostrum) or digestion.

After you have ensured that the tube is in the esophagus, the milk can be administered. Keep the head held up to allow gravity to aid

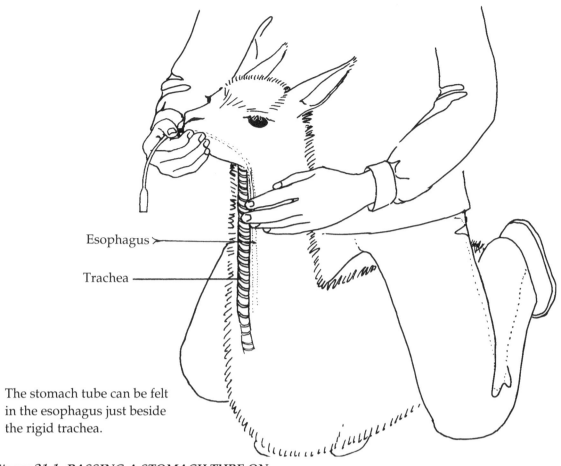

Esophagus

Trachea

The stomach tube can be felt
in the esophagus just beside
the rigid trachea.

*Figure 21.1 PASSING A STOMACH TUBE ON
A BABY LLAMA*

the milk's descent. A funnel, bottle, or dose syringe can be used to put the milk into the available end of the tube. It is important that the milk be given slowly, since the tube is not all the way to the stomach. If administered too fast, the milk would back up into the throat area and may be inhaled into the lungs. Care should be taken not to overfeed the milk. Mammalian stomachs can only hold 1/3 to 2/3 ounces per pound body weight of ingested food or drink.[1] To pull the tube out, place your thumb over the open end of the tube to prevent any milk from leaking out as you remove it, and slowly remove the tube.

Adult Llamas and Alpacas

Passing a stomach tube in an adult llama is more difficult and potentially more dangerous than in the baby. Adults tend to resist the procedure and regurgitate partially digested food which increases the likelihood of some aspiration of the food material into the lungs and subsequent life-threatening pneumonia. As with the newborn, if medications or food

are passed into the trachea rather than the esophagus, you can directly kill your llama. Therefore this procedure should NOT be done by llama owners, only by your veterinarian. The description below is intended for extenuating circumstances when your veterinarian may be unavailable, but has previously trained and advised you on the procedure.

To pass a stomach tube in an adult, restraint is extremely important. The llama needs to be in a chute and cross-tied to keep the head steady. Also, a mouth speculum, such as a strong piece of PVC pipe with the edges smoothed or a wooden block, or a metal tube, is necessary so that the llama cannot chew the tube. The stomach tube can be a horse stomach tube which is 1/2 inch outside diameter and 10 feet long. To get an approximate idea of how far you must pass the tube in order to reach the stomach, measure the tube from the mouth to the last rib. You can mark this distance with a marker on the tube. (See Figure 20.2.)

140

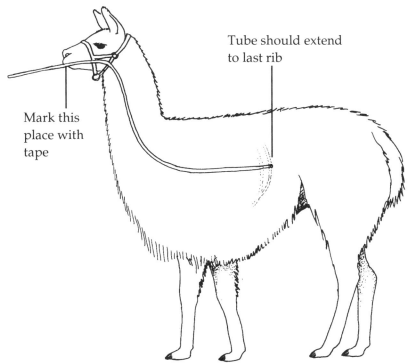

Tube should extend
to last rib

Mark this
place with
tape

Figure 21.2 MEASURING THE APPROPRIATE LENGTH OF TUBING FOR AN ADULT LLAMA.

Procedure:

First, lubricate the tip of the tube, then pass it through the speculum in the mouth, down the throat, and into the esophagus. (See Figure 20.3.) The tube should be felt moving just to the llama's left side of the trachea. (See Figure 20.4.) If placed improperly in the trachea, you will not feel the stomach tube with your fingers on the neck, as it will be dangling in the rigid trachea. Once in the stomach, you may hear or smell gas bubbles, like burps, from the tube. Sometimes blowing in the tube will elicit a burp. The odor from the llama's stomach contents smells like llama "spit". Once you are sure of the tube placement, administer the medicine as described above. (See Figure 20.5.)

Signs of Correct Stomach Tube Placement

1. Most important, feel for the stomach tube as it passes down the neck.

2. Resistance to passage. The esophagus is a collapsible tube that results in some resistance to passage of the tube.

3. A vacuum can be achieved by sucking on the end of the tube when it is in the esophagus. While doing this, the animal continues to breathe normally.

4. Stomach odor when the tube reaches the stomach.

Signs of Incorrect Stomach Tube Placement

1. Little resistance to passage.

2. Rattling of the tube when the trachea is shaken from side to side.

3. Absence of positive signs for esophageal passage.

4. Coughing or absence of coughing is an unreliable sign of the whereabouts of the tube!

Figure 21.3 PASSING A STOMACH TUBE IN AN ADULT LLAMA.

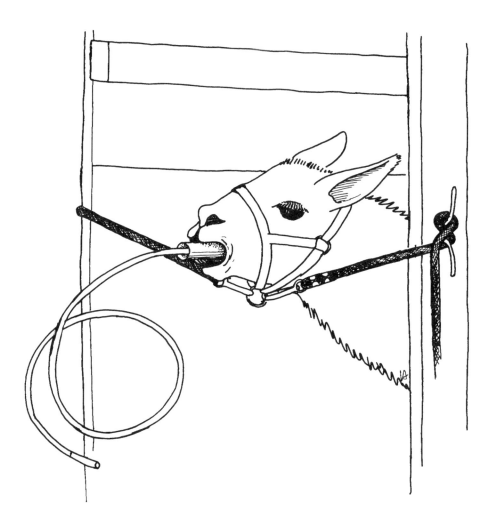

Caution: The tube and speculum must be held in place at all times to prevent the llama from swallowing it, or spitting out the speculum and chewing on the tube.

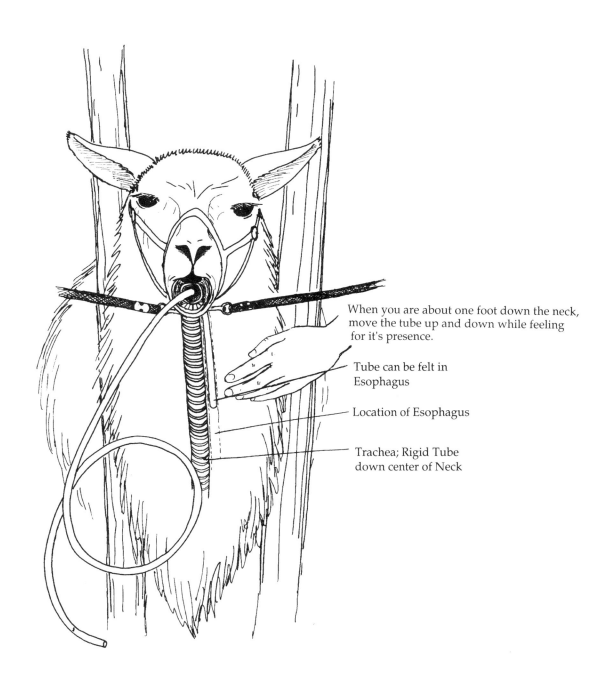

When you are about one foot down the neck, move the tube up and down while feeling for it's presence.

Tube can be felt in Esophagus

Location of Esophagus

Trachea; Rigid Tube down center of Neck

Figure 21.4 FEELING FOR CORRECT PLACEMENT OF TUBE.

Caution: The tube and speculum must be held in place at all times to prevent the llama from swallowing it, or spitting out the speculum and chewing on the tube.

Figure 21.5 ADMINISTERING MEDICINE THROUGH A STOMACH TUBE.

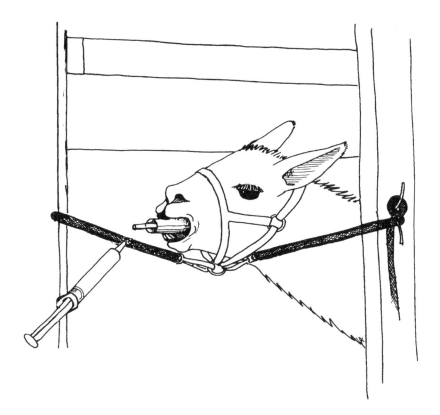

After giving the medicine, place your thumb over the end of the stomach tube and gently pull it out. After this procedure is complete, let the llama loose so that it won't regurgitate its medicine in anger.

Caution: The tube and speculum must be held in place at all times to prevent the llama from swallowing it, or spitting out the speculum and chewing on the tube.

CHAPTER 22:
TAKING A TEMPERATURE

Introduction

Taking a rectal temperature is one of the fundamentals of caring for llamas and alpacas. It allows for early diagnosis and treatment of many diseases. Sometimes, the body temperature rises when the only external signs of illness may be slight depression. Fever often precedes obvious signs of illness such as coughing, nasal discharge, and diarrhea. In addition, a rectal temperature is essential for the diagnosis of hypothermia and hyperthermia. It gives your veterinarian important information regarding whether or not your problem is an emergency. It is also essential in monitoring the progress of a sick animal.

Equipment

A large animal thermometer (5") works well for young adult and adult llamas. A smaller thermometer works better for the babies. In a pinch, a human thermometer may be used. However, once used on an animal, do not use it on humans. It is handy to tie a string on to the end of the thermometer before using it on the animal. This way, it can be clipped on to the hair of the tail or rump with a clothes pin or an alligator clip. This will save your thermometer from breaking when it falls out or is pushed out of the animal's rectum. Lubricating jelly is also needed for ease of sliding the thermometer into the rectum. Vaseline® or K-Y Jelly® works well.

Procedure

1. Shake the thermometer mercury down to below 95 degrees F.
2. Lift the llama's tail, and gently insert the lubricated thermometer into the anus. It should be inserted about 2 to 3 inches. Clip the string on to the hair.
3. Wait at least 3 minutes.
4. Unclip the string and remove the thermometer.
5. Wipe off the lubricant onto a towel and read the temperature.
6. Wipe off the thermometer with alcohol when you are finished.

Normal Adult Llama Temperature:
99 to 101.8 degrees F

Normal Adult Alpaca Temperature:
99.5 to 101.5 degrees F

**Normal Baby Llama and
Alpaca Temperature:**
100 to 102.2 degrees F

(Normal body temperatures vary with ambient temperature. Compare with other llamas or alpacas if it is elevated.)

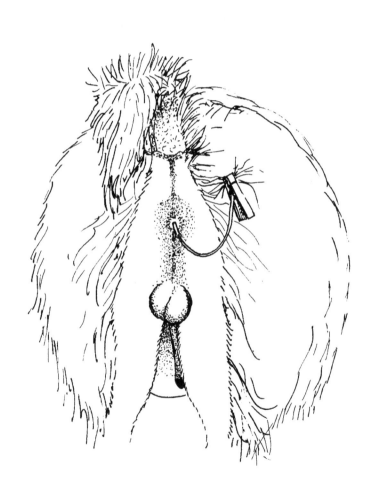

Figure 22.1 **TAKING A TEMPERATURE**

CHAPTER 23:

APPENDICES

Appendix I
Normal Values in Llamas and Alpacas[1]

- Normal life span 15 to 25 years in North America.
- Normal llama birth weight 18 to 40 pounds (30 pounds/13.6 Kg. average).
- Normal alpaca birth weight 8 to 23 pounds (16 to 20 pounds/7.3 to 9.1 Kg. average).
- Normal adult llama temperature 99 to 101.8 degrees F (37.2 to 38.7 degrees C).
- Normal adult alpaca temperature 93.5 to 101.5 degrees F. (37.5 to 38.6 degrees C).
- Normal baby (llama and alpaca) temperature 100 to 102.2 degrees F. (37.7 to 39 degrees C).
- Normal resting adult heart rate (llama and alpaca) 60 to 90 beats/minute.
- Normal baby llama heart rate 60 to 90 beats/minute.
- Normal baby alpaca heart rate 80 to 140 beats/minute.
- Normal resting llama respiration rate 10 to 30 breaths/minute.
- Normal resting alpaca respiration rate 20 to 30 breaths/minute.
- Normal adult llama weight 250 to 450 pounds (114 to 205 Kg.).
- Normal adult alpaca weight 110 to 185 pounds (50 to 84 Kg.).
- Normal ruminations (stomach gurgles) 2 to 4/minute.

Appendix II
Measurement Equivalents

1 fl oz = 30 mls = 30 cc

1 cc = 1 ml

1 pint = 480 mls = 16 fl. oz

1 liter 2.11 pints = 1.06 quarts 33.8 fl. oz

1 kilogram = 2.2 lbs

1 tsp = 5 mls

1 fl. oz = 2 TBSP

1 oz = 30 grams

1 gallon = 3800 mls = 3.8 liters

1 lb = 454 grams

1 grain = 60 mg

Appendix III
Supplies to Keep, on Hand: All Llamas and Alpacas

1. Pocket knife (it is a good idea to always carry one in your pocket)
2. 2 rectal thermometers (in case one breaks)
3. Epsom Salts
4. Toenail clippers
5. Enemas (Fleets®)
6. Spray bottles
7. Fly repellent

8. Sterile lubricant (K-Y Jelly®)

9. Cotton swabs (balls)

10. Isopropyl alcohol (rubbing alcohol)

11. Povidone iodine solution (Betadine® or Prepodyne® solution)

12. Povidone iodine scrub (Betadine® or Prepodyne® scrub). Scrub has soap in it so is better for removing dirt, but may be too harsh for tender areas.

13. 1 jar of antibiotic wound ointment such as Neosporin® or Betadine® ointments

14. 1 can of antibiotic wound spray

15. 1 puffer-bottle of antibiotic wound powder such as nitrofurazone

16. 1 to 2 tubes of antibiotic ophthalmic ointment (such as Neosporin®)

17. 2 liters of sterile saline solution

18. Needles for injection. Common sizes include 20g x 1"; 20g x 1-1/2"; 18g x 1-1/2"

19. Syringes. Common sizes include 3cc, 12cc, 20cc. A curved-tip syringe is nice to have on hand for flushing wounds.

20. Scissors (regular pair and bandage scissors)

21. Sterile Telfa® pads (3" x 4"; 4" x 4")

22. Kling® gauze rolls (3 to 4"; keep 4 to 5 rolls on hand)

23. 2" x 2" and/or 4" x 4" gauze squares

24. Vetrap® (2 to 4"; keep 4 to 5 rolls on hand)

25. 2 rolls of cotton

26. Elastikon® (2"; keep 3 to 4 rolls on hand)

Breeding Llamas and Alpacas

1. Several tubes of sterile lubricant (K-Y Jelly®)

2. Obstetrical plastic sleeves

3. 7% strong iodine or Nolvasan® and container for treating umbilical stump

4. Umbilical tape

5. Bulb syringe (can be used to aspirate fluid from a newborn's mouth and nostrils.)

6. C.M.T. kit

7. Puppy feeding tube, funnel, bottle and lamb nipples, and milk source for newborn emergencies.

The above products are available at the supermarket, drug store, or animal feed and supply stores. These products may also be conveniently purchased from your local veterinarian.

Restraint Chutes

Rocky Mountain Llama's
Versatile Llama Chute

*Designed and written
by Bobra Goldsmith*

The chute illustrated in the accompanying drawings has proven to be extremely easy to use for all the various medical and routine maintenance procedures necessary for our llama herds. For the most part we prefer to leave the sides open. There are occasions when we want one or both sides to be solid. Then we simply wire plywood sections cut to fit each side onto the horizontal poles. The plywood sides are 3/8" thick, and are stored inside or out of the weather when not in use.

In most cases, a simple cross-tie is adequate to restrain the llama. Cinches can be added as needed for less cooperative llamas.

A stanchion setup, an ad⌐ the principles of Stan Ebel's p⌐ chute, can be used for such pr⌐ passing a stomach tube in an adult ⌐ this device, the llama's head is pulleo the opening, the stanchion is closed ⌐ ⌐t his shoulders cannot pass through, arⱡd his head is fastened firmly so that he cannot move forward or backward. Care must be taken so that the llama does not back out of the stanchion while it is closed, or there is danger of his abrading the skin over the side of the eyes. The use of the stanchion offers the ultimate control of forward and backward movement by the llama, and good control of the head.

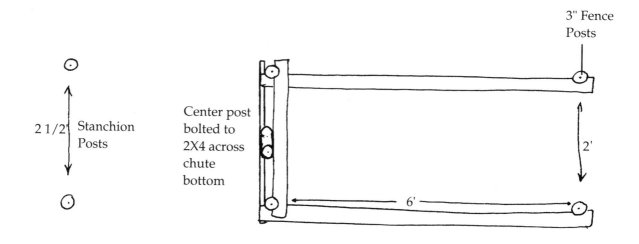

2 1/2' Stanchion Posts

Center post bolted to 2X4 across chute bottom

3" Fence Posts

2'

6'

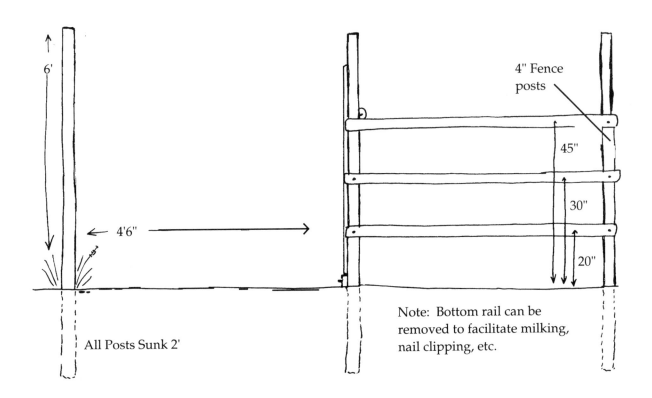

6'

4' Fence posts

45"

30"

20"

← 4'6" →

All Posts Sunk 2'

Note: Bottom rail can be removed to facilitate milking, nail clipping, etc.

Llama cross-tied in chute, without use of stanchion. (Adequate for most purposes.)

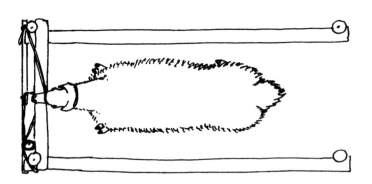

Rocky Mountain Llama's Versatile Llama Chute (cont'd)

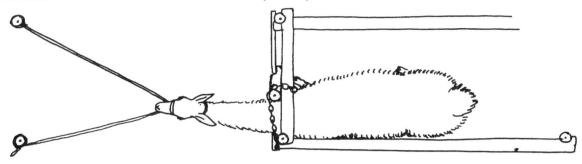

Llama tied in stanchion

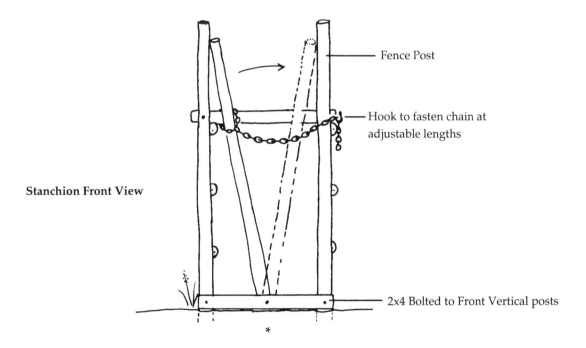

Stanchion Front View

— Fence Post

— Hook to fasten chain at adjustable lengths

— 2x4 Bolted to Front Vertical posts

*

*Detail

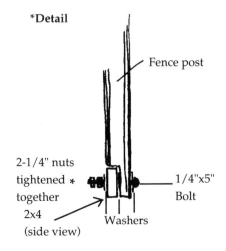

Fence post

2-1/4" nuts tightened * together

2x4 (side view)

Washers

1/4"x5" Bolt

This chute is not portable, but is inexpensive, easy to build, and as its name suggests, very versatile.

Stan Ebel's Llama Restraint Chute

Designed and written by Stan Ebel and Jim Hook of Great Divide Llamas, Loveland, Colo. (used with permission).

Materials

- 3/4" or 1" marine plywood *(for sides and floor)*
- 2" x 4" dimension lumber *(for frame)* two 8 ft. fence posts
- 4" top *(untreated or Pentatreated)*
- Hardware as described per figures and construction plans.

Construction

Frame per dimension in Figures:

1. Build the three upright sections separately. Use l/4" bolts (3-1/2" on all joints but the top of the middle section where 5" bolts are required) to secure the joints. Use washers at head and nut. The construction is adequate for most applications. If using on large numbers of animals, however, stability and longevity will be enhanced by adding metal angles on upper, outer corners of the frame (Figures 2 and 4). They're 1/8" plate with 8" sides. Additional 1/4" bolts are used to anchor points of angle pieces (1/4" x 2").

2. When uprights are complete, join them together with the 8 ft. cross members. Use 3-1/2" or 4" lag bolts to put these in place. Pre-drill holes of small diameter to prevent splitting. With these in place, put in cross member in back portion of floor as shown in Figure 1. Nail this in place with 16p small box nails.

3. Put in sides (4 ft. x 8 ft. plywood sheets). Nail in place with 16p small box or smaller ring shank nails. Next lay in floor and nail in place. Toenail sides to cross member in back floor section described in Step 2.

4. Cut off posts in vertical position, even with the top of the center frame section. Then with the posts in vertical position in the center of chute, mark 1/2" below the bottom of the top cross member of the center frame section. Then cut down with a saw from opposite sides so a center section approximately 1-1/4" in width is left. Use a chisel to cut away the sections sawed down.

Drill a 9/16" or 5/8" hole in the opposite end of the post to a depth of 4 to 5". Cut slots in the floor immediately behind the bottom cross member of the middle upright frame section. Make the

(Continued on page 151.)

FINISHED CHUTE

Stan Ebel's Llama Restraint Chute (cont'd)

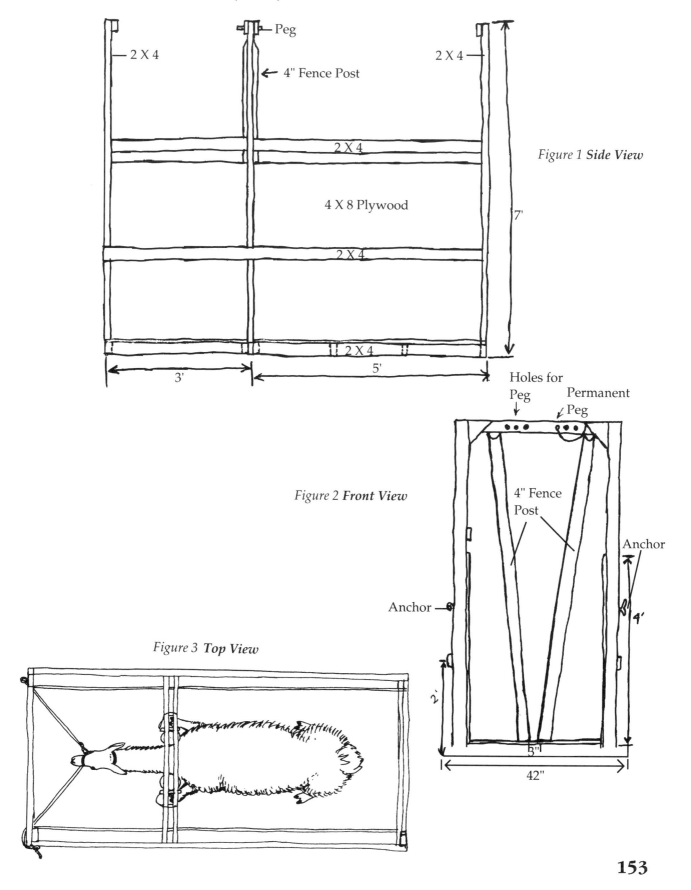

Figure 1 **Side View**

Figure 2 **Front View**

Figure 3 **Top View**

153

Figure 4
Front View

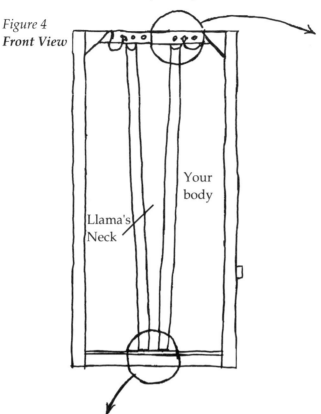

Your body

Llama's Neck

Stanchion Top Details

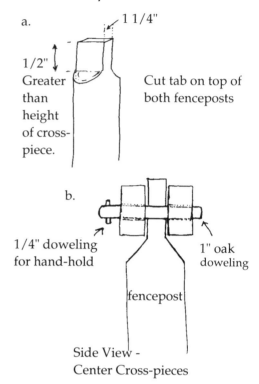

a.

1 1/4"

1/2"
Greater than height of cross-piece.

Cut tab on top of both fenceposts

b.

1/4" doweling for hand-hold

1" oak doweling

fencepost

Side View -
Center Cross-pieces

c.

Pegs may be attached to frame by a light chain or cord.

Stanchion Bottom Details

a.

Drill hole 9/16 on 5/8" diameter, 4-5" deep on bottom of fenceposts

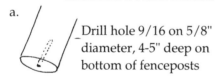

b.

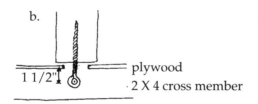

1 1/2"

plywood

2 X 4 cross member

Front View

Sideview

c.

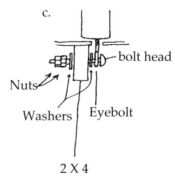

bolt head

Nuts

Washers   Eyebolt

2 X 4

154

slots 3/4" side x 1-1/2" long. Put posts in vertical position and space according to Figure 2 to locate these slots properly. Now tip the chute on its side.

Place post in position in the chute while it is on its side. Be sure to place tab in between the two horizontal cross members at top of frame. Now push 1/2" x 6" eyebolt through slot in floor and into hole drilled in bottom of post. Push eyebolt in until center of eye is 1-1/2" from top of the floor cross member. Mark this point and align it with center of floor slot. Drill hole in cross member. (Diameter of hole dictated by size of eye in eyebolt. Probably a 1/2" or larger bolt will work. Bolt must be long enough to allow washers and double nut).

Assemble per diagram. Leave loose enough to allow eyebolt free pivot motion around bolt. Tighten the two nuts against each other. Repeat for other post.

5. A series of 3 to 5 holes are drilled in the two cross pieces to accommodate 1" oak doweling pegs. They should be spaced to accommodate various options of human body size and llama size. Make sure the pegs and holes allow easy insertion and removal.

6. For head restraint, two ties are used. A permanent tie is established on one side (whichever fits your situation). An adjustable heavy web strap with panic snap on one end and heavy snap swivel on the other is ideal. Attach panic strap to lag bolt anchor (Figure 2) Use a boat mooring tie down on the other side to tie lead rope to.

The animal must be haltered to use the chute. A heavy nylon web halter is preferable. The animal is led in from the back (Figure 3). The posts may be in the open position (Figure 2) or pre-set to allow leader's body to pass to one side, while allowing llama's head through center. Close them enough so the shoulders will not clear (see Figure 3). Attach the quick-release tie snap to the halter ring. Then pull the animal with the lead so its shoulders are firmly against the posts and its neck is extended. Then tie the lead to the mooring anchor. The animal should not be able to move its head sideways or up and down. It should not be able to move forward or backward. If it lays down, breathing should not be hampered.

Do not allow the animal to lie on its side, as undue pressure will be placed on the trachea. Adjustment of the post assembly is important so it allows the shoulders to rest against the posts and they are not so tight as to put undue pressure on the neck. This is where the variable peg holes are necessary. It is also important to tie the assembly firmly to the side to minimize movement. It is best to use two people, especially on untrained animals. One controls the animal, the other the post assembly.

When releasing the animal, untie the lead first and allow enough slack to release permanent tie. Then fully release the lead to the control person and open post assembly so the llama can walk on through. Some larger animals may require backing out.

NOTE: This chute can be lifted into a pickup bed for transport and this may be facilitated by four strategically placed handles.

Appendix V
Congenital Defects in Llamas That are Likely Genetic in Origin Based on Observations in Other Species.

Head Abnormalities

1. Cataracts (eye lens opacity)
2. Choanal atresia (See Chapter 13)
3. Cleft palate (incomplete fusion of left and right sides of palate)
4. Coloboma of eyelids (cleft of eyelid)
5. Entropion (inversion of eyelid margin)
6. Hydrocephalus (abnormal accumulation of fluid in the cranium)
7. Palate agenesis (failure of palate formation)
8. Prognathism (shortened upper jaw) and other jaw deformities

Skeletal Abnormalities

1. Angular limb deformity (crooked legs)
2. Arthrogryposis (persistent contracture of a joint)
3. Erect pastern
4. Hemivertebrae (incomplete development of one side of one or more backbones)
5. Medial/lateral luxation of patella (kneecap dislocated towards/away from the midline)
6. Polydactylia (extra toes)
7. Spina bifida (defective closure of bone surrounding the spinal cord)
8. Syndactyly (fused toes)
9. Tail defects

Reproductive Abnormalities

1. Cryptorchidism/ectopic testiciles (retained testicles/abnormally located testicles)
2. Double cervix
3. Double uterus
4. Hermaphroditism (presence of male and female sex organs)
5. Hypospadia (abnormal development causing the urethra to open in an abnormal location)
6. Ovarian aplasia/ovarian hypoplasia, accompanied by uterine hypoplasia (no development/incomplete development of ovaries with incomplete development of the uterus)
7. Persistent hymen (imperforate hymen)
8. Pseudohermaphroditism (gonadal tissue present for one sex but has some characteristics of the opposite sex)
9. Segmental aplasia of female reproductive tract (lack of development of sections of genitals)
10. Segmental aplasia of seminiferous tubules (lack of development of regions in testicles where sperm are produced)
11. Shortened penis
12. Supranumerary teats (extra teats)
13. Testicular hypoplasia (incomplete testicular development)
14. Uterus unicornis (only one horn of the uterus develops instead of two)

Cardiovascular Abnormalities

1. Patent ductus arteriosus (persistence of an embryonic artery that bypasses the lungs)
2. Persistent truncus arteriosus (a single arterial trunk arising from the heart rather than one to the lungs and one to the rest of the body)
3. Ventricular septal defect (opening between the left and right ventricular heart chambers)
4. Atrial septal defect (opening between the left and right atrial heart chambers)
5. Transposition of Aorta and Pulmonary Artery[2] (two very important arteries carrying blood away from the heart)

Miscellaneous

1. Atresia ani (imperforate anus)
2. Single kidney (one instead of two)
3. Streaked hairlessness
4. Umbilical hernia (tissue protruding through an abnormal opening in the belly wall at the site of the umbilicus.)

Appendix VI
Poisonous Plants

The following is a partial list of poisonous plants commonly occurring in the United States. Although this list is based on other domesticated ruminants, there is no reason to believe that the toxic principles differ in llamas. Llamas tend to be browsers and alpacas are grazers. Both may be tempted to sample some of these poisonous plants. Prevention of ingestion of harmful plants includes avoiding over-grazing of pastures, preventing roadside eating on trips, and preventing ingestion of unknown plants on pack trips.

Plant Name	Distribution and Habitat	Remarks
Acacia greggi (Catclaw)	Southwestern Arizona. Limestone cliff areas and along streams.	Cyanide poisoning. Most poisonings in the fall.
Aconitum spp. (Monkshood)	Western United States. Moist Meadows.	Alkaloid poisoning.
Aesculus spp. (Buckeyes)	Eastern United States. Wooded areas.	Seeds and shoots of these trees or shrubs are especially poisonous. Most toxic in spring and summer.
Agave lecheguilla	Southwestern United States. Limestone hills and valleys.	Causes photosensitization.
Agrostemma githago (Corn cockle)	Throughout United States. Fields and waste areas.	Seeds are toxic.
Ambrosia tomentosa Nutt. (Bursage, White rag weed)	Four Corners area. Dry plains, waste ground, hills.	Nitrate poisoning. Grows in large dense patches.
Amianthium muscaetoxicum (Fly poison, Stagger grass, Crow poison)	Eastern United States. Fields and acid bogs.	Alkaloid poisoning.
Apocynum cannabinum L. (Dogbane)	Throughout the United States. Along streams and roadsides.	Resin and glycoside toxicity. Generally distasteful.
Asclepias spp. (Milkweed)	Many species covering most of the United States. Along old stream beds and irrigation canals.	Resinoid toxicity. Highly toxic. Dangerous in hay. Generally distasteful. Causes weakness, digestive upsets and convulsions.

157

Plant Name	Distribution and Habitat	Remarks
Astragalus spp. (Two-grooved milk vetch, Narrow-leaf poison vetch)	Rocky Mountain and neighboring states. Mid to lower elevations in dry meadows and hills in seleniferous soils.	Selenium toxicity and locoism. Plants are toxic year round.
Baccharis spp. (Silverling)	Eastern and Southeastern United States. Moist open areas.	This shrub is most toxic in the early spring and fall.
Brassica, Raphanus, Descurainia spp. (Mustards, Crucifers, Cress)	Throughout the United States. Fields and roadsides.	Fresh and dried plants are toxic. Mustard oil is toxic and can cause abortion.
Cerocarpus montanus Raf. (Mountain Mahogany)	Rocky Mountain region. Dry slopes and mesas.	Cyanide poisoning. Most poisonings occur in the fall.
Cicuta douglasii (DC.) (Water hemlock)	Western United States. Moist ground.	Highly toxic. Rapid death possible.
Conium maculatum (Poison hemlock)	Throughout the United States. Along roadsides and ditches.	Poison is an alkaloid called coniine. Ingestion of seeds is a common cause of poisoning.
Crotalaria spp.	Eastern and central United States. Roadsides and open areas.	Cumulative alkaloid toxicity. Dry plant also toxic. Often causes bloody diarrhea and death.
Datura spp. (Jimsonweed)	Southern and Western United States. Waste areas, trampled pastures, dry hills and plains.	Several toxic substances which can cause rapid death. Generally unpalatable.
Delphinium spp. (Larkspur)	Rocky Mountain regions. Plains, meadows and open woods.	Alkaloid toxicity. Most dangerous in the spring and early summer. Seeds are toxic in the fall.
Drymaria pachyphylla (Inkweed, Drymary)	Southwestern United States in dry alkaline soils of low areas.	Can cause diarrhea, coma and death.
Eupatorium rugosum (White snakeroot)	Eastern United States. Cleared areas and woods with rich soils.	Alcohol toxicity that may cause trembling, difficult breathing and death.

Plant Name	Distribution and Habitat	Remarks
Fagopyrum esculentum Moench (Buckwheat)	Scattered throughout the United States. Along roadsides.	Causes photosensitization.
Gelsemium sempervirens (Yellow jessamine)	Southeastern United States. Open woods and thickets.	Alkaloid toxicity.
Glottidium vesicarium, Sesbania spp., *Daubentonia punicea* (Bladder pod, Rattlebox, Sesbane)	Southeastern Coastal Plain states. Waste areas and open fields.	Saponin poisoning. Seeds are poisonous.
Grindelia spp. (Gum weed)	Central and Rocky Mountain states. Plains, prairies, and roadsides.	Selenium toxicity if growing in soil with high selenium.
Gutierrezia sarothrae (Broom Snakeweed)	Central and Western United States. Dry hills and plains.	Selenium toxicity if growing in soil with high selenium levels.
Halogeton glomeratus	Western United States, especially Utah, Colorado, Nevada, and Idaho. Dry alkaline soils.	Oxalate poisoning. Causes difficult breathing and death. Most frequent poisonings occur in the fall and winter.
Haplopappus heterophyllus (Rayless goldenrod, Burroweed)	Southwestern United States. Grasslands and plains.	Alcohol toxicity that may cause trembling, difficult breathing, and death.
Helenium hoopesii (Orange sneeze weed)	Rocky Mountain region. Moist and sunny areas.	Cumulative poison that causes vomiting and excessive salivation. Very dangerous in early spring and late fall.
Hymenoxys spp. (Rubberweed, Bitterweed)	Southwestern and western half of United States. Open ranges and roadsides.	Cumulative poison. Especially dangerous early spring and late fall. Generally distasteful.
Hyoscyamus niger (Black Henbane)	Southern Canada and Northern United States: some of the northern Rocky Mountain areas. Waste areas and dry roadside soil.	Rare poisoning. Distasteful to animals.

Plant Name	Distribution and Habitat	Remarks
Hypericum perforatum (St. Johnswort, Klamath weed)	Throughout the United States. Roadsides and waste areas.	Toxicity due to photosensitization. Toxic in fresh dried plants.
Iris missouriensis (Iris)	Western United States. Found in moist meadows.	Poisonous when fresh or dried. Moderately palatable.
Kalmia spp. (Laurel, Ivybush)	Northeastern and Eastern United States. Moist meadows and woods.	Causes salivation, vomiting, paralysis, and death.
Lantana spp.	Throughout the United States as an ornamental plant and wild in southern areas.	This shrub causes skin lesions and bloody diarrhea.
Lupinus argenteus (silvery lupine)	Montana to New Mexico and California. Dry hills and plains 4,000 to 11,000 feet in elevation.	Alkaloid toxicity. The seeds are the most toxic. Can cause congenital deformities.
Melia azedarach (Chinaberry)	Southeastern United States. Along fence rows and waste areas.	The fruit of this tree is most poisonous.
Nerium oleander (Oleander)	Southern United States. Domestic shrub or tree.	All parts of the plant are extremely toxic dry or green. Causes increased pulse rate, bloody diarrhea and death.
Nicotiana attenuata (Wild tobacco)	Western and Southwestern United States. Dry streambeds and flood plains.	Nicotine is the poison which causes nervous system problems. Generally distasteful.
Nolina texana (Beargrass)	Southwestern United States. Open hills.	Causes photosensitization. Flowers, fruits, and buds are toxic.
Notholaena sinuata var. cochisensis (Jimmy fern, Cloak fern)	Southwestern United States. Dry limestone rocky slopes.	Evergreen; causes abnormal gait. Most dangerous in fall and winter.
Oxytenia acerosa (Copperweed)	Southwestern United States. Desert ranges and foothills.	Plant increases in toxicity as it matures, thus most poisonous in fall and winter. Remains toxic when dry.

Plant Name	Distribution and Habitat	Remarks
Oxytropis spp. (Loco)	Scattered areas throughout United States. Low to middle elevations, hills, and dry slopes.	Causes locoism which attacks the motor and sensory nervous systems. Animals can develop a liking for these plants.
Phytolacca americana (Pokeweed)	Eastern United States in waste areas.	The roots are most poisonous.
Prunus spp. (Pin cherry, Wild red cherry, Western choke cherry)	Most of the United States. Canyons, woods, slopes and mountains.	Cyanide poisoning. Wilted leaves and bark very poisonous.
Pteridium aquilinum (Bracken fern)	Throughout the United States in dry or sandy areas.	Cumulative poison in fresh or dried plants.
Quercus spp. (Oak)	Most regions and habitats.	Gallotanin poisoning. Spring poisonings from young leaf buds and fall poisonings from dropped acorns.
Rhododendron spp.	Ornamental plant.	Can cause weakness, digestive upsets, muscle twitches.
Ricinus communes (Castor bean)	Present throughout United States as an ornamental plant and especially common in southern U.S.	Seeds are toxic (The poison is rendered harmless upon heating for castor oil preparation.)
Robinia spp. (Black locust)	Widely distributed along fence rows, around dwellings, along streams and valleys.	Bark from trees, sprouts, and trimmings are poisonous. Can cause diarrhea and shock.
Rumex crispus (Dock)	Throughout the United States in acidic or gravelly soils of pastures, meadows, and roadsides.	Poisonous if accumulates enough oxalates.
Salsola kali (Russian thistle)	Weed found especially in plains states. Dry disturbed areas.	Causes oxalate poisoning and nitrate poisoning if grown in heavily fertilized areas.

Plant Name	Distribution and Habitat	Remarks
Sarcobatus vermiculatus (Greasewood)	Plains states and west. Alkaline soils.	Oxalate poisoning. Maximum amount of oxalates in leaves in late summer. Must ingest large amounts over short periods of time.
Senecio spp. (Groundsel)	Species throughout the United States. Ranges, woods, prairies and plains.	Poisonous species due to alkaloids. A cumulative poison. Young plants are most toxic and dried plants remain toxic.
Solanum spp. (Nightshade, Jerusalem cherry, Buffalo burr)	Species throughout the United States. Woods, fence rows, open prairies, waste areas and roadsides.	Toxicity due to glycoalkaloids. Can cause weakness and vomiting.
Sophora secondiflora (mescal bean)	Southwestern United States. Canyons and hills in limestone soils.	Alkaloid toxicity affecting gait.
Sorghum halepense (Johnson grass) *Sorghum vulgare* (the cultivated sorghum)	Most of the United States except the extreme north. Along ditches and waste areas.	Toxic after cutting, trampling, drought or other stresses. Cyanide toxicity.
Stanleya pinnata (Princes's plume)	Rocky Mountain and plains states. Dry, desert, and seleniferous soils.	Selenium toxicity.
Suckleya suckleyana (Poison suckleys)	Montana to New Mexico. Drying water holes.	Cyanide toxicity. Palatable plant.
Taxus spp. (Yew)	Ornamental hedges throughout most of the United States.	Alkaloid toxicity. Causes gas, diarrhea, difficult breathing, collapse, death.
Tetradymia canescens (Horsebrush)	Rocky Mountain region and west to California. Dry hills, valleys and plains.	Cumulative poison causing photosensitization.

Plant Name	Distribution and Habitat	Remarks
Triglochin maritima (Arrow grass)	Most of the United States except for extreme west. Damp and marshy areas with alkaline soils.	Cyanide toxicity.
Veratrum spp. (False hellebore, Skunk cabbage)	Throughout United States in moist meadows, mountains, and valleys.	Alkaloid toxicity. Large amounts must be consumed for death, although small amounts may cause salivation, difficult breathing and decreased heart rate.
Xanthium spp. (Cocklebur)	Throughout the United States in fields and waste areas.	Most harmful in spring seedlings and also in older plants.
Zygadenus spp. (Death camas)	Throughout the United States especially in plains and western states. Depending on the species, they are found in moist plains, dry hills or dry plains.	Alkaloid toxicity. Plant bulbs are extremely toxic and dried plants are also poisonous. Can cause convulsions, congenital defects and abortion.

Merck, Sharp & Dohme Research Laboratories. *The Merck Veterinary Manual*, 5th edition. Rahway, NJ: Merck and Co., Inc., 1979, pp. 1019-1033.

Walter, R., Ph.D., "Poisonous Plants." Department of Botany and Plant Pathology, Colorado State University, Ft. Collins, CO, 1980, pp. 1-57.

Appendix VII
Resources on Fiber and Shearing

Appleyard, H.M., *Guide to the Identification of Animal Fibres*, 2nd Edition, (Leeds, England: WIRA, The Research and Services Centre for Textiles and Clothing, 1978).

The Fiberfest Magazine, P.O. Box 112, Dept. 18, Hastings, MI 49058.

Hoffman, Eric and Murray Fowler, D.V.M., *The Alpaca Book*, 1st Edition (Herald, CA: Clay Press, Inc., 1995).

ILA Brochure #9 *Llama Wool*, P.O. Box 37505, Denver, CO 80237.

Spin-Off Magazine, Interweave Press, 201 E. 4th Street, Loveland, CO 80537.

Switzer, Chris, *Spinning Llama and Alpaca*, Llamas' Store, P.O. Box 100, Herald, CA 95638.

Von Bergen, W., *Wool Handbook*, 3rd Edition, Volume 1, (New York: Interscience Publishers, John Wiley and Sons, 1963).

Yocum-McColl Testing Laboratories, Inc., 540 W. Elk Place, Denver, CO 80216-1823.

Appendix VIII
Feed Analysis Laboratories

Check with your local Extension Agent for a laboratory near you.

Colorado Analytical Lab; Post Office Drawer 507, Brighton, CO 80601; (303)659-2313.

Northeast DHIA Forage Lab; 730 Warren Road, Ithaca, NY 14850; (607)257-1272.

ENDNOTES

Chapter 2

1. Eric Hoffman and Murray Fowler, D.V.M., *The Alpaca Book,* 1st Edition (Herald, CA: Clay Press, Inc., 1995), p. 145.

Chapter 3

1. Hoffman and Fowler, *The Alpaca Book,* p. 122.

Chapter 4

1. Hoffman and Fowler, *The Alpaca Book,* p. 127.
2. L.W. Johnson, D.V.M., Ph.D., "Feeding Llamas," (ILA Educational Brochure, 1986), No. 6.
3. L.W. Johnson, D.V.M., Ph.D., *Llama Nutrition in Health and Disease* (Proceedings from RMLA Conference, Fall 1988), p. 4.
4. Murray Fowler, D.V.M., Ph.D., "Book Review," *Llamas Magazine,* Vol. 3, No. 6, p. 114.
5. Hoffman and Fowler, *The Alpaca Book,* p. 127.
6. L.W. Johnson, D.V.M., Ph.D., Proceedings from *Llama Medicine Workshop for Veterinarians,* 1988, p. 6.
7. Hoffman and Fowler, *The Alpaca Book,* p. 130.
8. Johnson, "Feeding Llamas."
9. Johnson, *Proceedings,* p. 7.
10. Murray Fowler, D.V.M., Ph.D., *Medicine and Surgeey of South American Camelids,* 1st Edition (Ames, IA: Iowa State University Press, 1989), p. 18.
11. Fowler, *Medicine,* pp. 17-18.
12. Hoffman and Fowler, *The Alpaca Book,* p. 127.
13. Johnson, *Proceedings,* p. 7.
14. Roland Smith, "Berserk Male Syndrome: Effects of Inappropriate Imprinting," *Llama Life,* Autumn 1987, pp. 3, 8.

Chapter 5

1. Virginia Ivens, et al., *Principal Parasites of Domestic Animals in the United States* (Illinois: The Board of Trustees of the University of Illinois, 1978).
2. Ivens.
3. Ivens.
4. Ivens.
5. Ivens.
6. John M. Cheney, D.V.M., "Paralaphostrongylus," *Proceedings from Llama Medicine Workshop for Veterinarians,* 1986, pp. 49-56.
7. Ivens.
8. Fowler, *Medicine,* p. 151.
9. Merck, Sharp & Dohme Research Laboratories, *The Merck Veterinary Manual,* 5th Edition (Rahway, NJ: Merck & Co., Inc., 1979), p. 415.
10. J. M. Hutchinson, B.V.Sc., "Ill-Thrift in Juvenile Llamas," *Proceedings from 52nd Annual Conference for Veterinarians,* 1991, p. 40.
11. Fowler, *Medicine,* pp. 134-135.
12. Cheney.
13. Murray Fowler, D.V.M., Ph.D., *Proceedings from Llama Medicine Workshop for Veterinarians,* 1987.
14. Erwin G. Pearson, et al., "Suspected chlorpyrifos toxicosis in llama, and plasma pseudocholinesterase activity in llamas given chlorpyrifos," *Journal of the American Veterinary Medical Association,* Vol. 198, No. 9), (Nov. 1, 1986), pp. 1062-1064.
15. R.B. Hillman, "Equine Disease Control Management," in *Animal Health and Nutrition,* May/June 1987, pp. 24-29.
16. Merck, Sharp & Dohme Research Laboratories, *The Merck Veterinary Manual,* 5th Edition, (Rahway, NJ: Merck & Co., Inc., 1979), pp. 396-409.
17. Philip D. Van Harreveld, et al., "Llama Castration," *Compendium,* April 2000, pp. 88-92.

Chapter 6

1. Murray Fowler, D.V.M., Ph.D., *Proceedings from Llama Medicine Workshop for Veterinarians,* 1986.

2. Murray Fowler, D.V.M., Ph.D., *Proceedings from Llama Medicine Workshop for Veterinarians,* 1987, p. 6.

Chapter 8

1. D.V. Hanselka, "Wounds and Their Management," *Equine Medicine and Surgery,* 3rd Edition, Vol. 2 (Santa Barbara, CA: American Veterinary Publications, 1982), pp. 857-858.

Chapter 9

1. Hoffman and Fowler, *The Alpaca Book,* p. 222.

Chapter 10

1. Fowler, *Medicine,* p. 136.

2. Rod A.W. Rosychuk, D.V.M., "The Llama Integumentary System in Health and Disease," *Proceedings from Llama Medicine Workshop for Veterinarians,* 1988, pp. 25-29.

3. Merck, Sharp & Dohme Research Laboratories, *The Merck Veterinary Manual,* 7th Edition (Rahway, NJ: Merck & Co., Inc., 1991), p. 344.

Chapter 12

1. Merck, *The Merck Veterinary Manual,* 5th Ed., (1979), p. 1483.

2. Fowler, *Medicine,* p. 170.

3. Merck, *The Merck Veterinary Manual,* 7th Ed., p. 1687.

Chapter 13

1. A.L. Kiorpes, D.V.M., Ph.D., et al., "Repair of Complete Choanal Atresia in a Llama," *Journal of the American Veterinary Medical Association,* Vol. 189, No. 9 (Nov. 1, 1986), pp. 1169-1171.

Chapter 14

1. Merck, *The Merck Veterinary Manual,* 7th Ed., p. 399.

Chapter 15

1. Fowler, *Medicine,* p. 246.

2. S.J. Tornquist, R.J. Van Saum, B.B. Smith, et al., "Hepatic lipidosis in llamas and alpacas: 31 cases (1991-1997), *Journal of the American Veterinary Medical Association,* 214: 1368-1372, 1999.

3. Colorado Department of Public Health and Environment. *West Nile Virus and Mosquito-Borne Viruses in Colorado.* 2003.

4. James G. Sikarski, D.V.M., M.S., *Notes on Meningeal Worm in Llamas.* (ILA Conference Notebook, Bellaire, MI., June 1988) pp. 29, 31.

Chapter 16

1. Johnson, *Proceedings,* p. 9.

2. Walter Bravo, "Breeding Practices in South America," *Llama Life,* Autumn 1988, pp. 16-17.

3. Bravo.

4. Johnson, *Proceedings,* p. 11.

5. Fowler, *Medicine,* p. 381.

6. Hoffman and Fowler, *The Alpaca Book,* p. 173.

Chapter 17

1. Johnson, *Proceedings,* p. 19.

2. Hoffman and Fowler, *The Alpaca Book,* p. 206.

3. Hoffman and Fowler, *The Alpaca Book,* p. 208.

4. Michael Strain, D.V.M., & Susan Strain, D.V.M., "Caring for the Premature Llamas," *Veterinary Medicine,* Dec. 1987, pp. 1243-1244.

Chapter 18

1. Fowler, *Medicine,* p. 189.

2. Fowler, *Medicine,* p. 190.

Chapter 21

1. Hoffman and Fowler, *The Alpaca Book,* p. 211.

Appendix

1. Hoffman and Fowler, *The Alpaca Book,* pp. 192, 206.

2. Fowler, *Medicine,* p. 357.

REFERENCES

Adams, O.R. *Lameness in Horses,* 3rd Edition Philadelphia, PA: Lea and Febiger, 1974.

Bishop, Janell and Lora Richard. "Fecal Survey of Llamas (Lama glama) in Oregon: Incidental Recovery of Nematodirus battus." *Journal of the American Veterinary Medical Association.* Vol. 191, No. 12 (Dec. 15, 1987), pp. 1579-1581.

Bravo, Waiter, D.V.M., "Breeding Practices in South America." *Llama Life,* Autumn 1988, pp. 16-17.

Foreyt, W.J. and John Lagerquist. "Experimental Infections of *Eimeria alpacae* and *Eimeria punoensis* in Llamas *(Lama glama). Journal of Parasitiology,* Vol. 78, No. 5, Oct. 1992, pp. 906-909.

Fowler, Murray, D.V.M., Ph.D. *Medicine and Surgery of South American Camelids,* 1st Edition (Ames, IA: Iowa State University Press, 1989).

Fowler, Murray, D.V.M., Ph.D., L.W. Johnson, D.V.M., Ph.D., et al. "Proceedings from Llama Medicine Workshop for Veterinarians." Colorado State University, Veterinary Teaching Hospital, l986, 1987, 1988.

Hanselka, D.V.M., "Wounds and Their Management," in *Equine Medicine and Surgery,* 3rd Edition, Vol. 2. Santa Barbara, CA: American Veterinary Publications, 1982.

Hillman, R.B., "Equine Disease Control Management." *Animal Health and Nutrition,* May/June 1987, pp. 24-29.

Hoffman, Eric and Murray Fowler, D.V.M. *The Alpaca Book,* 1st Edition (Herald, CA: Clay Press, Inc., 1995).

Ivens, Virginia, Daniel Mark and Normal Levine. *Principal Parasites of Domestic Animals in the United States.* Illinois: The Board of Trustees of the University of Illinois, 1978.

Johnson, L.W., D.V.M., Ph.D., "Feeding Llamas." International Llama Association Educational Brochure 6, 1986.

_____. *Llama Nutrition in Health and Disease,* Proceedings from RMLA Conference, Fall 1988.

Kiorpes, A.L., D.V.M., W.A. Lindsay, D.V.M., W.M. Adams, D.V.M., "Repair of Complete Choanal Atresia in a Llama." *Journal of the American Veterinary Medical Association,* Vol. 189, No. 9 (Nov. 1, 1986), pp. 1169-1171.

Kirkpatrick, Carl E., D.V.M., Ph.D., "Giardiasis in Large Animals." *Compendium on Continuing Education for the Practicing Veterinarian,* Vol. 11, No. 1 (Jan. 1989), pp. 80-84.

Massachusetts Medical Society. *Morbidity and Mortality Weekly Report,* Vol. 39, No. 12 (March 30, 1990), pp. 203-204.

Merck, Sharp & Dohme Research Laboratories. *The Merck Veterinary Manual,* 5th Edition. Rahway, NJ: Merck & Co., Inc. 1979.

Merck, Sharp & Dohme Research Laboratories. *The Merck Veterinary Manual,* 7th Edition. Rahway, NJ: Merck & Co., Inc. 1991.

Pearson, Erwin G., et al. "Suspected Chlorpyrifos Toxicosis in a Llama, and Plasma Pseudocholinesterase Activity in Llamas Given Chlorpyrifos." *Journal of the American Veterinary Medical Association.* Vol. 198, No. 9 (Nov. 1, 1986), pp. 1062-1064.

Silarskie, James G., D.V.M., M.S., *Notes on Meningeal Worm in Llamas.* ILA Conference Notebook, Bellaire, MI, June 1988.

Smith, Roland, "Berserk Male Syndrome: Effects of Inappropriate Imprinting." *Llama Life,* Autumn 1987, pp. 3, 8.

Strain, Michael, D.V.M., and Susan Strain, D.V.M., Caring for the Premature Llama. *Veterinary Medicine,* Dec. 1987, pp. 1243-1244.

Tornquist, S.J., R.J. Van Saum, B.B. Smith, et al., "Hepatic lipidosis in llamas and alpacas: 31 cases (1991-1997), *Journal of the American Veterinary Medical Association,* 214: 1368-1372, 1999.

Van Harreveld, Philip D., et al., "Llama Castration," *Compendium,* April 2000, pp. 88-92.

INDEX

Page numbers in *italics* indicate an illustration

Sutures, 58, 62, 65-67
Sweat glands, 73
Swelling, 50, 69-70
Syringes for flushing, 58, *59*, 65, 71, 148
    (see also Injections)

T

Tapeworm, 43, 48
Tearing (see Eyes)
Teats, 4, 16, 109, *110*, 123, *123*, 124-25, 156
Teeth, 41, 53-56, 91
    deciduous, baby, 53, 55
    fighting, 53, *53*, 54, *55*, 56, *56*, 65
    incisors, 16, 53, *53*, 54-55
Temperature
    alpacas, normal, 145, 147
    baby animals, 145, 147
    heat stroke, 85
    hypothermia, 86-87
    llamas, normal, 145, 147
    rectal measurement, 85-86, 145, *146*
Testicles, 16, 99, 156
Tetanus, 49-50, 65-66
Thelazia californiensis, 48, 60
Thermometer reading, 87, 145
Thread-necked strongyle, 42
Threadworm, 43
Tick paralysis, 45, 96
Ticks, 45-46, 48, 96, 98
Toenails
    torn, 81
    trimming, 75-81, *75-79*, 81
        (see also Feet, legs)
Torn ears, 65
Toxemia, 96, 98
Toxic substances, 92 (see also Poisonous,
    toxic plants)
Toxoplasmosis, 108-09
Trachea, 90, 139-40, *140*, 141, *143*
    (see also Stomach tube)
Tracheostomy, 90
Trailers, 20-21

Transportation
    down llamas, 97
    health concerns, 19, 20-22, 60, 83, 86
    vehicles, methods, 19-21
Trichostrongylus (stomach worm), 42, 48
Trichuris (whipworm), 42-43, 48
Tuberculosis, 90
    tests, 17, 19
Tumors, 70, 89
Twins, 108

U

Ulcers, 60, 70, 96
Ultrasound exams, 16, 105-07
Umbilical cord/umbilicus, 110, 114, 116, 122
Umbilical hernia, 16, 156
Uncoordinated feet, legs, 44-45, 97
    (see also Feet, legs)
Undercoat, 73
Undulant fever (see Brucellosis)
Urinary obstruction, 96
Uterine
    biopsy, culture, 107
    horns, 100, 107
    infections, 96, 107
    torsions, 117
Uterus, 100, 156

V

Vaccination
    boosters, 49-50, 65-67, 120
    program, 41, 49-50, 94
Vaccines, 49-50, 65-66
Vagina, 100
Vaginal discharge, 107
Valley Fever, 74
Vans, 21-22
Ventilation, 21-22, 51, 90
Vesicular stomatitis, 91
Virus diseases, 50, 91
Vitamins, 34-35
Vulva, 16, 100, *105*, 106-07, 109, *111*, 116-17

ABOUT THE PUBLISHER

RMLA
Rocky Mountain Llama and Alpaca Association

The Rocky Mountain Llama Association began when a group of about 40 enthusiasts met in 1982 at Monument, Colorado, to join together in an effort to promote educating the general public and one another about the wonders of llamas. Among the founders were individuals interested in the use of llamas for their fiber and for packing; persons who enjoyed their companionship as pets; and serious breeders interested in breeding genetically sound animals.

The founders were pioneers in promoting soundness and conformational strengths of llamas, sharing experiences of all uses of llamas, and stimulating the beginnings of research. The cornerstone of a great organization was well established by these farsighted organizers. RMLA received its incorporation documents from the State of Colorado in late 1983. Today, the organization solidly continues in the direction set by our founding members for the purpose of sharing information about llamas and alpacas with fellow owners and the general public.

- The RMLA's Member's Lending Library is one of the largest collection of materials assimilated about llamas and alpacas and contains old and out-of-print publications as well as the very latest publications and videos.

- RMLA sponsors either a major conference touching on many new and old issues OR a mini-conference which combines with an event pertinent to the industry.

- Many special events are sponsored to introduce llamas and alpacas to the public. One can attend the Fairplay Race, see the Fiber Co-op at such places as the Taos Wool Festival and the Estes Park Wool Market, show or visit in an organized show or attend and have fun at many other events - large or small.

- RMLA and its members worked as the Canyonlands Task Force and has managed the Llama Legal Defense Fund. This action has proven that llamas do not pose a disease threat to wildlife on public lands. RMLA was a major contributor to research that has proven that llamas are environmentally friendly.

- RMLA sponsors many ALSA Llama and Alpaca Shows.

Some of RMLA's Activities and Benefits:

- The Journal of the RMLA has grown into one of North America's largest and most informative llama and alpaca publications...a great place to read about the personal experience of fellow owners and new information available concerning your animals.

- A portion of your membership dues, net sales from Caring for Llamas and Alpacas, and other monetary gifts are accrued in the RMLA Research Fund.

- RMLA Fiber Co-op is a wonderful way for its members to market their animals' fiber. Each year, thousands of dollars worth of fiber and fiber products are sold through the co-op.

RMLA on the Internet

RMLA has its own presence on the Internet. People all over the world can check the calendar of events and read past articles from the Journal. All library books and videos are listed. There is even a section dedicated to our Service Directory and a Fiber Room. Most recently a page has been added so that new members can join RMLA on-line and past members can renew their membership.

www.RMLA.com

ROCKY MOUNTAIN LLAMA AND ALPACA ASSOCIATION, INC.

Recognizes these Life Members who have dedicated themselves to the betterment of llamas and alpacas by financially supporting the RMLA

1. Betty Robertson, Englewood, CO
2. Bob & Barbara Hance, Wheat Ridge, CO
3. Pat & Maury Cox, DVM, LaSalle, CO
4. Ron & Lougene Baird, Sedalia, CO
5. Judy Sealy, Loma, CO
6. John & Sharon Beacham, Salida, CO
7. Marie & Richard Hoover, Elizabeth, CO
8. Richard & Jeanne Williams, Kemmerer, WY
9. Mark Baker, Gracemont, OK
10. Gue Grimm, Georgetown, CO
11. Suzanne Hartung, Steamboat Springs, CO
12. Robert Mallonee, Albuquerque, NM
13. Bruce & Gay Ellis, Franktown, CO
14. Jerry Dunn, Golden, CO
15. John & Jane Stout, Glade Park, CO
16. Jan Cummer, Woodland Park, CO
17. Sandy & Sharon Pierce, La Vernia, TX
18. Regis & Gloria Thoma, Butler, PA
19. Sandra & David Lockwood, Confier, CO
20. Sondra Grumbein, Benton City, WA
21. Julie Heggie & Gayle Woodsum, Laramie, WY
22. Jane & Arthur Levene, Denver, CO
23. Bobra Goldsmith, Longmont, CO
24. Jan & Sherry Adamcyk, Kiowa, CO
25. Linda Kutscher & James Lewis, Berthoud, CO
26. Jim & Marta Haas, Evergreen CO
27. Paul & Karen Schwartz, Big Horn, WY

You are invited to Join RMLA

Membership year - April 1 through March 31

Life Membership: $500. Up to 2 adults per membership. Special recognition in Membership Directory, 10% discount on conference fees, one time free 1/4 page in the Journal of RMLA. (If there are any future special assessment, Life Members Pay those the same as annual members would.)

Annual Membership: $40 per year. Allows up to 2 (two) adults to vote. Provideds a ranch/farm listing on the RMLA website.

Youth Membership: $10 per year. Requires an adult-members sponsor. For young llama lovers under 19 years of age.

Young Adult Membership: $10 per year. Requires a parent-member sponsor. For the former Youth member only, who are students beyond High School. 1 vote.

Only RMLA members have access to our Lending Library

All members who keep their e-mail address current with RMLA Membership Committee are placed on the RMLA e-mail Newsline and receive many timely announcements and items of interest to all RMLA members.

All members receive Directories: Journals and/or Newsletters, one copy pers household; when they are published; & urgent or timely llama related e-mail messages, provided you give us your e-mail address.

Visit our website at **www.RMLA.com**